RANDOM
HOUSE
LARGE
PRINT

Bad Blood

Bad Blood

Secrets and Lies in a Silicon Valley Startup

JOHN CARREYROU

RANDOM HOUSE
LARGE PRINT

Published in the United States of America by Random House Large Print in association with Alfred A. Knopf, a division of Penguin Random House LLC, New York.

Cover design by Tyler Comrie

The Library of Congress has established a Cataloging-in-Publication record for this title.

ISBN: 978-1-9848-3363-1

www.penguinrandomhouse.com/large-print-format-books

FIRST LARGE PRINT EDITION

Printed in the United States of America

10 9 8 7 6 5 4

This Large Print edition published in accord with the standards of the N.A.V.H.

For Molly, Sebastian, Jack, and Francesca

Contents

Author's Note

This book is based on hundreds of interviews with more than 150 people, including more than sixty former Theranos employees. Most of the men and women who appear as characters in the narrative do so under their real names, but some asked that I shield their identities, either because they feared retribution from the company, worried that they might be swept up in the Justice Department's ongoing criminal investigation, or wanted to guard their privacy. In the interest of getting the most complete and detailed rendering of the facts, I agreed to give these people pseudonyms. However, everything else I describe about them and their experiences is factual and true.

Any quotes I have used from emails or documents

are verbatim and based on the documents themselves. When I have attributed quotes to characters in dialogues, those quotes are reconstructed from participants' memories. Some chapters rely on records from legal proceedings, such as deposition testimony. When that's the case, I have identified those records at length in the notes section at the end of the narrative.

In the process of writing this book, I reached out to all of the key figures in the Theranos saga and offered them the opportunity to comment on any allegations concerning them. Elizabeth Holmes, as is her right, declined my interview requests and chose not to cooperate with this account.

Bad Blood

Prologue

November 17, 2006

Tim Kemp had good news for his team.

The former IBM executive was in charge of bioinformatics at Theranos, a startup with a cutting-edge blood-testing system. The company had just completed its first big live demonstration for a pharmaceutical company. Elizabeth Holmes, Theranos's twenty-two-year-old founder, had flown to Switzerland and shown off the system's capabilities to executives at Novartis, the European drug giant.

"Elizabeth called me this morning," Kemp wrote in an email to his fifteen-person team. "She expressed her thanks and said that, 'it was perfect!'

She specifically asked me to thank you and let you all know her appreciation. She additionally mentioned that Novartis was so impressed that they have asked for a proposal and have expressed interest in a financial arrangement for a project. We did what we came to do!"

This was a pivotal moment for Theranos. The three-year-old startup had progressed from an ambitious idea Holmes had dreamed up in her Stanford dorm room to an actual product a huge multinational corporation was interested in using.

Word of the demo's success made its way upstairs to the second floor, where senior executives' offices were located.

One of those executives was Henry Mosley, Theranos's chief financial officer. Mosley had joined Theranos eight months earlier, in March 2006. A rumpled dresser with piercing green eyes and a laid-back personality, he was a veteran of Silicon Valley's technology scene. After growing up in the Washington, D.C., area and getting his MBA at the University of Utah, he'd come out to California in the late 1970s and never left. His first job was at chipmaker Intel, one of the Valley's pioneers. He'd later gone on to run the finance departments of four different tech companies, taking two of them public. Theranos was far from his first rodeo.

What had drawn Mosley to Theranos was the talent and experience gathered around Elizabeth. She might be young, but she was surrounded by an all-star cast. The chairman of her board was Donald L. Lucas, the venture capitalist who had groomed billionaire software entrepreneur Larry Ellison and helped him take Oracle Corporation public in the mid-1980s. Lucas and Ellison had both put some of their own money into Theranos.

Another board member with a sterling reputation was Channing Robertson, the associate dean of Stanford's School of Engineering. Robertson was one of the stars of the Stanford faculty. His expert testimony about the addictive properties of cigarettes had forced the tobacco industry to enter into a landmark $6.5 billion settlement with the state of Minnesota in the late 1990s. Based on the few interactions Mosley had had with him, it was clear Robertson thought the world of Elizabeth.

Theranos also had a strong management team. Kemp had spent thirty years at IBM. Diane Parks, Theranos's chief commercial officer, had twenty-five years of experience at pharmaceutical and biotechnology companies. John Howard, the senior vice president for products, had overseen Panasonic's chip-making subsidiary. It wasn't often that you found executives of that caliber at a small startup.

It wasn't just the board and the executive team that had sold Mosley on Theranos, though. The market it was going after was huge. Pharmaceutical companies spent tens of billions of dollars on clinical trials to test new drugs each year. If Theranos could make itself indispensable to them and capture a fraction of that spending, it could make a killing.

Elizabeth had asked him to put together some financial projections she could show investors. The first set of numbers he'd come up with hadn't been to her liking, so he'd revised them upward. He was a little uncomfortable with the revised numbers, but he figured they were in the realm of the plausible if the company executed perfectly. Besides, the venture capitalists startups courted for funding knew that startup founders overstated these forecasts. It was part of the game. VCs even had a term for it: the hockey-stick forecast. It showed revenue stagnating for a few years and then magically shooting up in a straight line.

The one thing Mosley wasn't sure he completely understood was how the Theranos technology worked. When prospective investors came by, he took them to see Shaunak Roy, Theranos's cofounder. Shaunak had a Ph.D. in chemical engineering. He and Elizabeth had worked together in Robertson's research lab at Stanford.

Shaunak would prick his finger and milk a few drops of blood from it. Then he would transfer the blood to a white plastic cartridge the size of a credit card. The cartridge would slot into a rectangular box the size of a toaster. The box was called a reader. It extracted a data signal from the cartridge and beamed it wirelessly to a server that analyzed the data and beamed back a result. That was the gist of it.

When Shaunak demonstrated the system to investors, he pointed them to a computer screen that showed the blood flowing through the cartridge inside the reader. Mosley didn't really grasp the physics or chemistries at play. But that wasn't his role. He was the finance guy. As long as the system showed a result, he was happy. And it always did.

ELIZABETH WAS BACK from Switzerland a few days later. She sauntered around with a smile on her face, more evidence that the trip had gone well, Mosley figured. Not that that was unusual. Elizabeth was often upbeat. She had an entrepreneur's boundless optimism. She liked to use the term "**extra**-ordinary," with "extra" written in italics and a hyphen for emphasis, to describe the Theranos mission in her emails to staff. It was a bit over the top, but she seemed sincere and Mosley knew that evangelizing was what successful startup founders did in Silicon Valley. You didn't change the world by being cynical.

What was odd, though, was that the handful of colleagues who'd accompanied Elizabeth on the trip didn't seem to share her enthusiasm. Some of them looked outright downcast.

Did someone's puppy get run over? Mosley wondered half jokingly.

He wandered downstairs, where most of the company's sixty employees sat in clusters of cubicles, and looked for Shaunak. Surely Shaunak would know if there was any problem he hadn't been told about.

At first, Shaunak professed not to know anything. But Mosley sensed he was holding back and kept pressing him. Shaunak gradually let down his guard and allowed that the Theranos 1.0, as Elizabeth had christened the blood-testing system, didn't always work. It was kind of a crapshoot, actually, he said. Sometimes you could coax a result from it and sometimes you couldn't.

This was news to Mosley. He thought the system was reliable. Didn't it always seem to work when investors came to view it?

Well, there was a reason it always **seemed** to work, Shaunak said. The image on the computer screen showing the blood flowing through the cartridge and settling into the little wells was real. But you never knew whether you were going to get a result or not. So they'd recorded a result from one of the times it worked. It was that recorded result that was displayed at the end of each demo.

Mosley was stunned. He thought the results were extracted in real time from the blood inside the cartridge. That was certainly what the investors he brought by were led to believe. What Shaunak had just described sounded like a sham. It was

OK to be optimistic and aspirational when you pitched investors, but there was a line not to cross. And this, in Mosley's view, crossed it.

So, what exactly had happened with Novartis?

Mosley couldn't get a straight answer from anyone, but he now suspected some similar sleight of hand. And he was right. One of the two readers Elizabeth took to Switzerland had malfunctioned when they got there. The employees she brought with her had stayed up all night trying to get it to work. To mask the problem during the demo the next morning, Tim Kemp's team in California had beamed over a fake result.

MOSLEY HAD a weekly meeting with Elizabeth scheduled for that afternoon. When he entered her office, he was immediately reminded of her charisma. She had the presence of someone much older than she was. The way she trained her big blue eyes on you without blinking made you feel like the center of the world. It was almost hypnotic. Her voice added to the mesmerizing effect: she spoke in an unusually deep baritone.

Mosley decided to let the meeting run its natural course before bringing up his concerns. Theranos had just closed its third round of funding. By any measure, it was a resounding success: the company had raised another $32 million from investors, on top of the $15 million raised in its first two funding rounds. The most impressive number was its new valuation: **one hundred and sixty-five million dollars**. There weren't many three-year-old start-ups that could say they were worth that much.

One big reason for the rich valuation was the agreements Theranos told investors it had reached with pharmaceutical partners. A slide deck listed six deals with five companies that would generate revenues of $120 million to $300 million over the next eighteen months. It listed another fifteen deals under negotiation. If those came to fruition, revenues could eventually reach $1.5 billion, according to the PowerPoint presentation.

The pharmaceutical companies were going to use Theranos's blood-testing system to monitor patients' response to new drugs. The cartridges and readers would be placed in patients' homes during clinical trials. Patients would prick their fingers several times a day and the readers would beam their blood-test results to the trial's sponsor. If the results indicated a bad reaction to the drug, the drug's maker would be able to lower the dosage immediately rather than wait until the end of the trial. This would reduce pharmaceutical companies' research costs by as much as 30 percent. Or so the slide deck said.

Mosley's unease with all these claims had grown since that morning's discovery. For one thing, in his eight months at Theranos, he'd never laid eyes on the pharmaceutical contracts. Every time he inquired about them, he was told they were "under legal review." More important, he'd agreed to those ambitious revenue forecasts because he thought the Theranos system worked reliably.

If Elizabeth shared any of these misgivings, she showed no signs of it. She was the picture of a relaxed and happy leader. The new valuation, in particular, was a source of great pride. New directors might join the board to reflect the growing roster of investors, she told him.

Mosley saw an opening to broach the trip to Switzerland and the office rumors that something had gone wrong. When he did, Elizabeth admitted that there had been a problem, but she shrugged it off. It would easily be fixed, she said.

Mosley was dubious given what he now knew. He brought up what Shaunak had told him about the investor demos. They should stop doing them if they weren't completely real, he said. "We've been fooling investors. We can't keep doing that."

Elizabeth's expression suddenly changed. Her cheerful demeanor of just moments ago vanished and gave way to a mask of hostility. It was like a switch had been flipped. She leveled a cold stare at her chief financial officer.

"Henry, you're not a team player," she said in an icy tone. "I think you should leave right now."

There was no mistaking what had just happened. Elizabeth wasn't merely asking him to get out of her office. She was telling him to leave the company—immediately. Mosley had just been fired.

| ONE |

A Purposeful Life

Elizabeth Anne Holmes knew she wanted to be a successful entrepreneur from a young age.

When she was seven, she set out to design a time machine and filled up a notebook with detailed engineering drawings.

When she was nine or ten, one of her relatives asked her at a family gathering the question every boy and girl is asked sooner or later: "What do you want to do when you grow up?"

Without skipping a beat, Elizabeth replied, "I want to be a billionaire."

"Wouldn't you rather be president?" the relative asked.

"No, the president will marry me because I'll have a billion dollars."

These weren't the idle words of a child. Elizabeth uttered them with the utmost seriousness and determination, according to a family member who witnessed the scene.

Elizabeth's ambition was nurtured by her parents. Christian and Noel Holmes had high expectations for their daughter rooted in a distinguished family history.

On her father's side, she was descended from Charles Louis Fleischmann, a Hungarian immigrant who founded a thriving business known as the Fleischmann Yeast Company. Its remarkable success turned the Fleischmanns into one of the wealthiest families in America at the turn of the twentieth century.

Bettie Fleischmann, Charles's daughter, married her father's Danish physician, Dr. Christian Holmes. He was Elizabeth's great-great-grandfather. Aided by the political and business connections of his wife's wealthy family, Dr. Holmes established Cincinnati General Hospital and the University of Cincinnati's medical school. So the case could be made—and it would in fact be made to the venture capitalists clustered on Sand Hill Road near the Stanford University campus—that Elizabeth didn't just inherit entrepreneurial genes, but medical ones too.

Elizabeth's mother, Noel, had her own proud family background. Her father was a West Point graduate who planned and carried out the shift from a draft-based military to an all-volunteer force as a high-ranking Pentagon official in the early 1970s. The Daousts traced their ancestry all the way back to the **maréchal** Davout, one of Napoleon's top field generals.

But it was the accomplishments of Elizabeth's father's side of the family that burned brightest and captured the imagination. Chris Holmes made sure to school his daughter not just in the outsized success of its older generations but also in the failings of its younger ones. Both his father and grandfather had lived large but flawed lives, cycling through marriages and struggling with alcoholism. Chris blamed them for squandering the family fortune.

"I grew up with those stories about greatness," Elizabeth would tell **The New Yorker** in an interview years later, "and about people deciding not to spend their lives on something purposeful, and what happens to them when they make that choice—the impact on character and quality of life."

ELIZABETH'S EARLY YEARS were spent in Washington, D.C., where her father held a succession of jobs at government agencies ranging from the State Department to the Agency for International Development. Her mother worked as an aide on Capitol Hill until she interrupted her career to raise Elizabeth and her younger brother, Christian.

During the summers, Noel and the children

headed down to Boca Raton, Florida, where Elizabeth's aunt and uncle, Elizabeth and Ron Dietz, owned a condo with a beautiful view of the Intracoastal Waterway. Their son, David, was three and a half years younger than Elizabeth and a year and a half younger than Christian.

The cousins slept on foam mattresses on the condo's floor and dashed off to the beach in the mornings for a swim. The afternoons were whiled away playing Monopoly. When Elizabeth was ahead, which was most of the time, she would insist on playing on to the bitter end, piling on the houses and hotels for as long as it took for David and Christian to go broke. When she occasionally lost, she stormed off in a fury and, more than once, ran right through the screen of the condo's front door. It was an early glimpse of her intense competitive streak.

In high school, Elizabeth wasn't part of the popular crowd. By then, her father had moved the family to Houston to take a job at the conglomerate Tenneco. The Holmes children attended St. John's, Houston's most prestigious private school. A gangly teenage girl with big blue eyes, Elizabeth bleached her hair in an attempt to fit in and struggled with an eating disorder.

During her sophomore year, she threw herself into her schoolwork, often staying up late at night to study, and became a straight-A student. It was the start of a lifelong pattern: work hard and sleep little. As she excelled academically, she also managed to find her footing socially and dated the son of a respected Houston orthopedic surgeon. They traveled to New York together to celebrate the new millennium in Times Square.

As college drew closer, Elizabeth set her sights on Stanford. It was the obvious choice for an accomplished student interested in science and computers who dreamed of becoming an entrepreneur. The little agricultural college founded by railroad tycoon Leland Stanford at the end of the nineteenth century had become inextricably linked with Silicon Valley. The internet boom was in full swing then and some of its biggest stars, like Yahoo, had been founded on the Stanford campus. In Elizabeth's senior year, two Stanford Ph.D. students were beginning to attract attention with another little startup called Google.

Elizabeth already knew Stanford well. Her family had lived in Woodside, California, a few miles from the Stanford campus, for several years in the late 1980s and early 1990s. While there, she had become friends with a girl who lived next door named Jesse Draper. Jesse's father was Tim Draper, a third-generation venture capitalist who was on his way to becoming one of the Valley's most successful startup investors.

Elizabeth had another connection to Stanford: Chinese. Her father had traveled to China a lot for work and decided his children should learn Mandarin, so he and Noel had arranged for a tutor to come to the house in Houston on Saturday mornings. Midway through high school, Elizabeth talked her way into Stanford's summer Mandarin program. It was only supposed to be open to college students, but she impressed the program's director enough with her fluency that he made an exception. The first five weeks were taught on the Stanford campus in Palo Alto, followed by four weeks of instruction in Beijing.

ELIZABETH WAS ACCEPTED to Stanford in the spring of 2002 as a President's Scholar, a distinction bestowed on top students that came with a three-thousand-dollar grant she could use to pursue any intellectual interest of her choosing.

Her father had drilled into her the notion that she should live a purposeful life. During his career in public service, Chris Holmes had overseen humanitarian efforts like the 1980 Mariel boatlift, in which more than one hundred thousand Cubans and Haitians migrated to the United States. There were pictures around the house of him providing disaster relief in war-torn countries. The message Elizabeth took away from them is that if she wanted to truly leave her mark on the world, she would need to accomplish something that furthered the greater good, not just become rich. Biotechnology offered the prospect of achieving both. She chose to study chemical engineering, a field that provided a natural gateway to the industry.

The face of Stanford's chemical engineering department was Channing Robertson. Charismatic, handsome, and funny, Robertson had been teaching at the university since 1970 and had a rare ability to connect with his students. He was also by far the hippest member of the engineering faculty, sporting a graying blond mane and showing up to class in leather jackets that made him seem a decade younger than his fifty-nine years.

Elizabeth took Robertson's Introduction to Chemical Engineering class and a seminar he taught on controlled drug-delivery devices. She also lobbied him to let her help out in his research lab. Robertson agreed and farmed her out to a Ph.D. student who was working on a project to find the best enzymes to put in laundry detergent.

Outside of the long hours she put in at the lab, Elizabeth led an active social life. She attended campus parties and dated a sophomore named JT Batson. Batson was from a small town in Georgia and was struck by how polished and worldly Elizabeth was, though he also found her guarded. "She wasn't the biggest sharer in the world," he recalls. "She played things close to the vest."

Over winter break of her freshman year, Elizabeth returned to Houston to celebrate the holidays with her parents and the Dietzes, who flew down from Indianapolis. She'd only been in college for a few months, but she was already entertaining thoughts of dropping out. During Christmas dinner, her father floated a paper airplane toward her end of the table with the letters "P.H.D." written on its wings.

Elizabeth's response was blunt, according to a family member in attendance: "No, Dad, I'm not interested in getting a Ph.D., I want to make money."

That spring, she showed up one day at the door of Batson's dorm room and told him she couldn't see him anymore because she was starting a company and would have to devote all her time to it. Batson, who had never been dumped before, was stunned but remembers that the unusual reason she gave took some of the sting out of the rejection.

Elizabeth didn't actually drop out of Stanford until the following fall after returning from a summer internship at the Genome Institute of Singapore. Asia had been ravaged earlier in 2003 by the spread of a previously unknown illness called severe acute respiratory syndrome, or SARS, and Elizabeth had spent the summer testing patient specimens obtained with old low-tech methods like syringes and nasal swabs. The experience left her convinced there must be a better way.

When she got back home to Houston, she sat down at her computer for five straight days, sleeping one or two hours a night and eating from trays of food her mother brought her. Drawing from new technologies she had learned about during her internship and in Robertson's classes, she wrote a patent application for an arm patch that would simultaneously diagnose medical conditions and treat them.

Elizabeth caught up on sleep in the family car while her mother drove her from Texas to California to start her sophomore year. As soon as she was back on campus, she showed Robertson and Shaunak Roy, the Ph.D. student she was assisting in his lab, her proposed patent.

In court testimony years later, Robertson recalled being impressed by her inventiveness: "She had somehow been able to take and synthesize these pieces of science and engineering and technology in ways that I had never thought of." He was also struck by how motivated and determined she was to see her idea through. "I never encountered a student like this before of the then thousands of students that I had talked" to, he said. "I encouraged her to go out and pursue her dream."

Shaunak was more skeptical. Raised by Indian immigrant parents in Chicago, far from the razzle-dazzle of Silicon Valley, he considered himself very pragmatic and grounded. Elizabeth's concept seemed to him a bit far-fetched. But he got swept up in Robertson's enthusiasm and in the notion of launching a startup.

While Elizabeth filed the paperwork to start a company, Shaunak completed the last semester of work he needed to get his degree. In May 2004, he joined the startup as its first employee and was granted a minority stake in the business. Robertson, for his part, joined the company's board as an adviser.

AT FIRST, Elizabeth and Shaunak holed up in a tiny office in Burlingame for a few months until they found a bigger space. The new location was far from glamorous. While its address was technically in Menlo Park, it was in a gritty industrial zone on the edge of East Palo Alto, where shootings remained frequent. One morning, Elizabeth showed up at work with shards of glass in her hair. Someone had shot at her car and shattered the driver's-side window, missing her head by inches.

Elizabeth incorporated the company as Real-Time Cures, which an unfortunate typo turned into "Real-Time Curses" on early employees' paychecks. She later changed the name to Theranos, a combination of the words "therapy" and "diagnosis."

To raise the money she needed, she leveraged her family connections. She convinced Tim Draper, the father of her childhood friend and former neighbor Jesse Draper, to invest $1 million. The Draper name carried a lot of weight and helped give Elizabeth some credibility: Tim's grandfather had founded Silicon Valley's first venture capital firm in the late 1950s, and Tim's own firm, DFJ, was known for lucrative early investments in companies like the web-based email service Hotmail.

Another family connection she tapped for a large investment, the retired corporate turnaround specialist Victor Palmieri, was a longtime friend of her father's. The two had met in the late 1970s during the Carter administration when Chris Holmes worked at the State Department and Palmieri served as its ambassador at large for refugee affairs.

Elizabeth impressed Draper and Palmieri with her bubbly energy and her vision of applying principles of nano- and microtechnology to the field of diagnostics. In a twenty-six-page document she used to recruit investors, she described an adhesive patch that would draw blood painlessly through the skin using microneedles. The TheraPatch, as the document called it, would contain a microchip sensing system that would analyze the blood and make "a process control decision" about how much of a drug to deliver. It would also communicate its readings wirelessly to a patient's doctor. The document included a colored diagram of the patch and its various components.

Not everyone bought the pitch. One morning in July 2004, Elizabeth met with MedVenture Associates, a venture capital firm that specialized in medical technology investments. Sitting across a conference room table from the firm's five partners, she spoke quickly and in grand terms about the potential her technology had to change mankind.

But when the MedVenture partners asked for more specifics about her microchip system and how it would differ from one that had already been developed and commercialized by a company called Abaxis, she got visibly flustered and the meeting grew tense. Unable to answer the partners' probing technical questions, she got up after about an hour and left in a huff.

MedVenture Associates wasn't the only venture capital firm to turn down the nineteen-year-old college dropout. But that didn't stop Elizabeth from raising a total of nearly $6 million by the end of 2004 from a grab bag of investors. In addition to Draper and Palmieri, she secured investments from an aging venture capitalist named John Bryan and from Stephen L. Feinberg, a real estate and private equity investor who was on the board of Houston's MD Anderson Cancer Center. She also persuaded a fellow Stanford student named Michael Chang, whose family controlled a multibillion-dollar distributor of high-tech devices in Taiwan, to invest. Several members of the extended Holmes family, including Noel Holmes's sister, Elizabeth Dietz, chipped in too.

As the money flowed in, it became apparent to Shaunak that a little patch that could do all the things Elizabeth wanted it to do bordered on science fiction. It might be theoretically possible, just like manned flights to Mars were theoretically possible. But the [illegible]vas in the details. In an attempt to make the patch concept more feasible, they pared it down to just the diagnostic part, but even that was incredibly challenging.

Eventually they jettisoned the patch altogether in favor of something akin to the handheld devices used to monitor blood-glucose levels in diabetes patients. Elizabeth wanted the Theranos device to be portable like those glucose monitors, but she wanted it to measure many more substances in the blood than just sugar, which would make it a lot more complex and therefore bulkier.

The compromise was a cartridge-and-reader system that blended the fields of microfluidics and biochemistry. The patient would prick her finger to draw a small sample of blood and place it in a cartridge that looked like a thick credit card. The cartridge would slot into a bigger machine called a reader. Pumps inside the reader would push the blood through tiny channels in the cartridge and into little wells coated with proteins known as antibodies. On its way to the wells, a filter would separate the blood's solid elements, its red and white blood cells, from the plasma and let only the plasma through. When the plasma came into contact with the antibodies, a chemical reaction would produce a signal that would be "read" by the reader and translated into a result.

Elizabeth envisioned placing the cartridges and readers in patients' homes so that they could test their blood regularly. A cellular antenna on the reader would send the test results to the computer of a patient's doctor by way of a central server. This would allow the doctor to make adjustments to the patient's medication quickly, rather than waiting for the patient to go get his blood tested at a blood-draw center or during his next office visit.

By late 2005, eighteen months after he'd come on board, Shaunak was beginning to feel like they were making progress. The company had a prototype, dubbed the Theranos 1.0, and had grown to two dozen employees. It also had a business model it hoped would quickly generate revenues: it planned to license its blood-testing technology to pharmaceutical companies to help them catch adverse drug reactions during clinical trials.

Their little enterprise was even beginning to attract some buzz. On Christmas Day, Elizabeth sent employees an email with the subject line "Happy Happy Holidays." It wished them well and referred them to an interview she had given to the technology magazine **Red Herring**. The email ended with, "And Heres to 'the hottest start-up in the valley'!!!"

| TWO |

The Gluebot

Edmond Ku interviewed with Elizabeth Holmes in early 2006 and was instantly captivated by the vision she unspooled before him.

She described a world in which drugs would be minutely tailored to individuals thanks to Theranos's blood-monitoring technology. To illustrate her point, she cited Celebrex, a painkiller that was under a cloud because it was thought to increase the risk of heart attacks and strokes. There was talk that its maker, Pfizer, would have to pull it from the market. With the Theranos system, Celebrex's side effects could be eliminated, allowing millions of arthritis sufferers to keep taking the drug to alleviate their aches and pains, she explained. Elizabeth cited the fact that an estimated one hundred thousand Americans died each year from adverse drug reactions. Theranos would eliminate all those deaths, she said. It would quite literally save lives.

Edmond, who went by Ed, felt himself drawn in by the young woman sitting across from him who was staring at him intently without blinking. The mission she was describing was admirable, he thought.

Ed was a quiet engineer who had gained a reputation in the Valley as a fix-it man. Tech startups stymied by a complex engineering problem called him and, more often than not, he found a solution. Born in Hong Kong, he had emigrated to Canada with his family in his early teens and had the habit common among native Chinese speakers who learn English as a second language of always speaking in the present tense.

A member of Theranos's board had recently approached him about taking over engineering at the startup. If he accepted the job, his task would be to turn the Theranos 1.0 prototype into a viable product the company could commercialize. After hearing Elizabeth's inspiring pitch, he decided to sign on.

It didn't take Ed long to realize that Theranos was the toughest engineering challenge he'd ever tackled. His experience was in electronics, not medical devices. And the prototype he'd inherited didn't really work. It was more like a mock-up of what

Elizabeth had in mind. He had to turn the mock-up into a functioning device.

The main difficulty stemmed from Elizabeth's insistence that they use very little blood. She'd inherited from her mother a phobia of needles; Noel Holmes fainted at the mere sight of a syringe. Elizabeth wanted the Theranos technology to work with just a drop of blood pricked from the tip of a finger. She was so fixated on the idea that she got upset when an employee bought red Hershey's Kisses and put the Theranos logo on them for a company display at a job fair. The Hershey's Kisses were meant to represent drops of blood, but Elizabeth felt they were much too big to convey the tiny volumes she had in mind.

Her obsession with miniaturization extended to the cartridge. She wanted it to fit in the palm of a hand, further complicating Ed's task. He and his team spent months reengineering it, but they never reached a point where they could reliably reproduce the same test results from the same blood samples.

The quantity of blood they were allowed to work with was so small that it had to be diluted with a saline solution to create more volume. That made what would otherwise have been relatively routine chemistry work a lot more challenging.

Adding another level of complexity, blood and saline weren't the only fluids that had to flow through the cartridge. The reactions that occurred when the blood reached the little wells required chemicals known as reagents. Those were stored in separate chambers.

All these fluids needed to flow through the cartridge in a meticulously choreographed sequence, so the cartridge contained little valves that opened and shut at precise intervals. Ed and his engineers tinkered with the design and the timing of the valves and the speed at which the various fluids were pumped through the cartridge.

Another problem was preventing all those fluids from leaking and contaminating one another. They tried changing the shape, length, and orientation of the tiny channels in the cartridge to minimize the contamination. They ran countless tests with food coloring to see where the different colors went and where the contamination occurred.

It was a complicated, interconnected system compressed into a small space. One of Ed's engineers had an analogy for it: it was like a web of rubber bands. Pulling on one would inevitably stretch several of the others.

Each cartridge cost upward of two hundred dollars to make and could only be used once. They were testing hundreds of them a week. Elizabeth had purchased a $2 million automated packaging line in anticipation of the day they could start shipping them, but that day seemed far off. Having already blown through its first $6 million, Theranos had raised another $9 million in a second funding round to replenish its coffers.

The chemistry work was handled by a separate group made up of biochemists. The collaboration between that group and Ed's group was far from optimal. Both reported up to Elizabeth but weren't encouraged to communicate with each other. Elizabeth liked to keep information compartmentalized so that only she had the full picture of the system's development.

As a result, Ed wasn't sure if the problems they were encountering were due to the microfluidics he was responsible for or the chemistry work he had nothing to do with. He knew one thing, though: they'd have a much better chance of success if Elizabeth allowed them to use more blood. But she wouldn't hear of it.

ED WAS WORKING late one evening when Elizabeth came by his workspace. She was frustrated with the pace of their progress and wanted to run the engineering department twenty-four hours a day, seven days a week, to accelerate development. Ed thought that was a terrible idea. His team was working long hours as it was.

He had noticed that employee turnover at the company was already high and that it wasn't confined to the rank and file. Top executives didn't seem to last long either. Henry Mosley, the chief financial officer, had disappeared one day. There was a rumor circulating around the office that he'd been caught embezzling funds. No one knew if there was any truth to it because his departure, like all the others, wasn't announced or explained. It made for an unnerving work environment: a colleague might be there one day and gone the next and you had no idea why.

Ed pushed back against Elizabeth's proposal. Even if he instituted shifts, a round-the-clock schedule would make his engineers burn out, he told her.

"I don't care. We can change people in and out," she responded. "The company is all that matters."

Ed didn't think she meant it to sound as callous as it did. But she was so laser focused on achieving her goals that she seemed oblivious to the practical implications of her decisions. Ed had noticed a quote on her desk cut out from a recent press article about Theranos. It was from Channing Robertson, the Stanford professor who was on the company's board.

The quote read, "You start to realize you are looking in the eyes of another Bill Gates, or Steve Jobs."

That was a high bar to set for herself, Ed thought. Then again, if there was anyone who could clear it, it might just be this young woman. Ed had never encountered anyone as driven and relentless. She slept four hours a night and popped chocolate-coated coffee beans throughout the day to inject herself with caffeine. He tried to tell her to get more sleep and to live a healthier lifestyle, but she brushed him off.

As obstinate as Elizabeth was, Ed knew there was one person who had her ear: a mysterious man named Sunny. Elizabeth had dropped his name enough times that Ed had gleaned some basic facts about him: he was Indian, he was older than Elizabeth, and they were a couple. The story was that Sunny had made a fortune from the sale of an internet company he'd cofounded in the late 1990s.

Sunny wasn't a visible presence at Theranos but he seemed to loom large in Elizabeth's life. At the company Christmas party in a Palo Alto restaurant in late 2006, Elizabeth got too tipsy to go home on her own, so she called Sunny and asked him to come pick her up. That's when Ed learned that they were living together in a condo a few blocks away.

Sunny wasn't the only older man giving Elizabeth advice. She had brunch with Don Lucas every Sunday at his home in Atherton, the ultrawealthy enclave north of Palo Alto. Larry Ellison, whom she'd met through Lucas, was also an influence. Lucas and Ellison had both invested in Theranos's second funding round, which in Silicon Valley parlance was known as a "Series B" round. Ellison sometimes dropped by in his red Porsche to check on his investment. It wasn't uncommon to hear Elizabeth start a sentence with "Larry says."

Ellison might be one of the richest people in the world, with a net worth of some $25 billion, but he wasn't necessarily the ideal role model. In Oracle's early years, he had famously exaggerated his database software's capabilities and shipped versions of it crawling with bugs. That's not something you could do with a medical device.

It was hard to know how much Elizabeth's approach to running Theranos was her own and how much she was channeling Ellison, Lucas, or Sunny, but one thing was clear: she wasn't happy when Ed refused to make his engineering group run 24/7. From that moment on, their relationship cooled.

Before long, Ed noticed that Elizabeth was making new engineering hires, but she wasn't having them report to him. They formed a separate group. A rival group. It dawned on him that she was pitting his engineering team and the new team against each other in some corporate version of survival of the fittest.

Ed didn't have time to dwell on it too much because there was something else he had to deal with: Elizabeth had convinced Pfizer to try out the Theranos system in a pilot project in Tennessee. Under the agreement, Theranos 1.0 units were going to be placed in people's homes and patients were going to test their blood with them every day. The results would be sent wirelessly to Theranos's office in California, where they would be analyzed and then forwarded to Pfizer. They had to somehow fix all the problems before the study started. She'd already scheduled a trip to Tennessee to begin training some of the patients and doctors in how to use the system.

In early August 2007, Ed accompanied Elizabeth to Nashville. Sunny picked them up from the office in his Porsche and drove them to the airport. It was the first time Ed met him in person. The extent of their age gap suddenly became apparent. Sunny looked to be in his early forties, nearly twenty years older than Elizabeth. There was also a cold, businesslike dynamic to their relationship. When they parted at the airport, Sunny didn't say "Goodbye" or "Have a nice trip." Instead, he barked, "Now go make some money!"

When they got to Tennessee, the cartridges and the readers they'd brought weren't functioning properly, so Ed had to spend the night disassembling and reassembling them on his bed in his hotel room. He managed to get them working well enough by morning that they were able to draw blood samples from two patients and a half dozen doctors and nurses at a local oncology clinic.

The patients looked very sick. Ed learned that they were dying of cancer. They were taking drugs designed to slow the growth of their tumors, which might buy them a few more months to live.

On their return to California, Elizabeth pronounced the trip a success and sent one of her cheerful emails to the staff.

"It was truly awesome," she wrote. "The patients grasped onto the system immediately. The minute you meet them you sense their fear, their hope, and their pain."

Theranos employees, she added, should "take a victory lap."

Ed didn't feel as upbeat. Using the Theranos 1.0 in a patient study seemed premature, especially now that he knew the study involved terminal cancer patients.

TO BLOW OFF STEAM, Ed went out for beers with Shaunak on Friday evenings at a raucous sports bar called the Old Pro in Palo Alto. Often, Gary Frenzel, the head of the chemistry team, would join them.

Gary was a good old boy from Texas. He liked to tell war stories about his days as a rodeo rider. He'd given up riding and pursued a career as a chemist after breaking too many bones. Gary loved to gossip and crack jokes, causing Shaunak to burst into a loud, high-pitched giggle that was the most ridiculous laugh Ed had ever heard. The three bonded during these outings and became good friends.

Then one day, Gary stopped coming to the Old Pro. Ed and Shaunak weren't sure why at first but they soon had their answer.

In late August 2007, an email went out to Theranos employees to gather upstairs for a meeting. The company had grown to more than seventy people. Everyone stopped what he or she was doing and assembled in front of Elizabeth's office on the second floor.

The mood was serious. Elizabeth had a frown on her face. She looked angry. Standing next to her was Michael Esquivel, a sharply dressed, fast-talking lawyer who had joined Theranos a few months earlier as its general counsel from Wilson Sonsini Goodrich & Rosati, Silicon Valley's premier law firm.

Esquivel did most of the talking. He said Theranos was suing three former employees for stealing its intellectual property. Their names were Michael O'Connell, Chris Todd, and John Howard. Howard had overseen all research and development and interviewed Ed before he was hired. Todd was Ed's predecessor and had led the design of the 1.0 prototype. And O'Connell was an employee who had worked on the 1.0 cartridge until he left the previous summer.

No one was to have any contact with them going forward and all emails and documents must be preserved, Esquivel instructed. He would be conducting a thorough investigation to gather evidence with the assistance of Wilson Sonsini. Then he added something that sent a jolt through the room.

"We've called the FBI to assist us with the case."

Ed and Shaunak figured Gary Frenzel was probably freaked out by this turn of events. He was good friends with Chris Todd, Ed's predecessor. Gary had worked with Todd for five years at two previous companies before following him to Theranos. After Todd had left Theranos in July 2006, he and Gary had remained in frequent contact, talking often on the phone and exchanging emails. Elizabeth and Esquivel must have found out and read Gary the riot act. He looked spooked.

Shaunak had been friendly with Todd too and was able to quietly piece together what had happened.

O'Connell, who had a postdoctorate in nanotechnology from Stanford, thought he had solved the microfluidic problems that hampered the Theranos system and had talked Todd into forming a company with him. They'd called it Avidnostics. O'Connell also held discussions with Howard, who'd provided some help and advice but declined to join their venture. Avidnostics was very similar to Theranos, except they planned on marketing their machine to veterinarians on the theory that regulatory approvals would be easier to obtain for a device that performed blood tests on animals rather than humans.

They'd pitched a few VCs, unsuccessfully, at which point O'Connell had lost patience and emailed Elizabeth to ask her if she wanted to license their technology.

Big mistake.

Elizabeth had always worried about proprietary company information leaking out, to an extent that sometimes felt overblown. She required not just employees to sign nondisclosure agreements, but anyone else who entered Theranos's offices or did business with it. Even within the company, she kept tight control over the flow of information.

O'Connell's actions confirmed her worst suspicions. Within days, she was laying the groundwork for a lawsuit. Theranos filed its fourteen-page complaint in California Superior Court on August 27, 2007. It requested that the court issue a temporary restraining order against the three former employees, appoint a special master "to ensure that they do not use or disclose Plaintiff's trade secrets," and award Theranos five different types of monetary damages.

In the ensuing weeks and months, the atmosphere at the office became oppressive. Document retention emails landed in employees' in-boxes with regularity and Theranos went into lockdown. The head of IT, a computer technician named Matt Bissel, deployed security features that made everyone feel under surveillance. You couldn't put a USB drive into an office computer without Bissel knowing about it. One employee got caught doing just that and was fired.

AMID THE DRAMA, the competition between engineering teams intensified. The new group competing with Ed's was headed by Tony Nugent. Tony was a gruff, no-nonsense Irishman who'd spent eleven years at Logitech, the maker of computer accessories, followed by a stint at a company called Cholestech that made a simpler version of what Theranos was trying to build. Its handheld product, the Cholestech LDX, could perform three cholesterol tests and a glucose test on small samples of blood drawn from a finger.

Tony had initially been brought to Theranos as a consultant by Gary Hewett, Cholestech's founder. He'd had to step into Hewett's shoes when Hewett was fired after just five months as Theranos's vice president of research and development.

Hewett's conviction when he'd arrived at Theranos was that microfluidics didn't work in blood diagnostics because the volumes were too small to allow for accurate measurements. But he hadn't had time to come up with much of an alternative. That job now fell to Tony.

Tony decided that part of the Theranos value proposition should be to automate all the steps that bench chemists followed when they tested blood in a laboratory. In order to automate, Tony needed a robot. But he didn't want to waste time building one from scratch, so he ordered a three-thousand-dollar glue-dispensing robot from a company in New Jersey called Fisnar. It became the heart of the new Theranos system.

The Fisnar robot was a pretty rudimentary piece of machinery. It was a mechanical arm fixed to a gantry that had three degrees of motion: right and left; forward and back; and up and down. Tony fastened a pipette—a slender translucent tube used to transfer or measure out small quantities of liquid—to the robot and programmed it to make the movements that a chemist would make in the lab.

With the help of another recently hired engineer named Dave Nelson, he eventually built a smaller version of the glue robot that fit inside an aluminum box a little wider and a little shorter than a desktop computer tower. Tony and Dave borrowed some components from the 1.0, like the electronics and the software, and added them to their box, which became the new reader.

The new cartridge was a tray containing little plastic tubes and two pipette tips. Like its microfluidic predecessor, it could only be used once. You placed the blood sample in one of the tubes and pushed the cartridge into the reader through a little door that swung upward. The reader's robotic arm then went to work, replicating the human chemist's steps.

First, it grabbed one of the two pipette tips and used it to aspirate the blood and mix it with diluents contained in the cartridge's other tubes. Then it grabbed the other pipette tip and aspirated the diluted blood with it. This second tip was coated with antibodies, which attached themselves to the molecule of interest, creating a microscopic sandwich.

The robot's last step was to aspirate reagents from yet another tube in the cartridge. When the reagents came into contact with the microscopic sandwiches, a chemical reaction occurred that emitted a light signal. An instrument inside the reader called a photomultiplier tube then translated the light signal into an electrical current.

The molecule's concentration in the blood—what the test sought to measure—could be inferred from the power of the electrical current, which was proportional to the intensity of the light.

This blood-testing technique was known as a chemiluminescent immunoassay. (In laboratory speak, the word "assay" is synonymous with "blood test.") The technique was not new: it had been pioneered in the early 1980s by a professor at Cardiff University. But Tony had automated it inside a machine that, though bigger than the toaster-size Theranos 1.0, was still small enough to make Elizabeth's vision of placing it in patients' homes possible. And it only required about 50 microliters of blood. That was more than the 10 microliters Elizabeth initially insisted upon, but it still amounted to just a drop.

By September 2007, four months after he'd started building it, Tony had a functioning prototype. One that performed far more reliably than the balky system Ed Ku was still laboring on in another part of the office.

Tony asked Elizabeth what she wanted to call it.

"We tried everything else and it failed, so let's call it the Edison," she said.

What some employees had taken to derisively calling the "gluebot" was suddenly the new way forward. And it now had a far more respectable name, inspired by the man widely considered to be America's greatest inventor.

The decision to abandon the microfluidic system in favor of the Edison was ironic given that Theranos had just filed a lawsuit to protect the intellectual property underpinning the former. It was also bad news for Ed Ku.

One morning a few weeks before Thanksgiving, Ed and his engineers were called into a conference room one after the other. When it was Ed's turn, Tony, a human resources manager named Tara Lencioni, and the lawyer Michael Esquivel informed him that he was being let go. The company was heading in a new direction and it didn't involve what he was working on, they said. Ed would have to sign a new nondisclosure and nondisparagement agreement if he wanted to get his severance. Lencioni and Esquivel walked him to his workspace to retrieve a few personal belongings and then escorted him out of the building.

About an hour later, Tony glanced out the window and noticed that Ed was still standing outside, his jacket slung over his arm, looking lost. It turned out he hadn't driven his car to the office that morning and was stranded. This was before the days of Uber, so Tony went to find Shaunak and, knowing that they were friends, asked him to drive Ed home.

Shaunak followed Ed out the door two weeks later, albeit on friendlier terms. The Edison was at its core a converted glue robot and that was a pretty big step down from the lofty vision Elizabeth had originally sold him on. He was also unsettled by the constant staff turnover and the lawsuit hysteria. After about three and a half years, it felt like time to move on. Shaunak told Elizabeth he was thinking of going back to school and they agreed to part ways. She organized an office party to see him off.

Theranos's product might no longer be the groundbreaking, futuristic technology she'd envisioned, but Elizabeth remained as committed as ever to her company. In fact, she was so excited about the Edison that she started taking it out of the office almost immediately to show it off. Tony quipped to Dave that they should have built two before telling her about it.

Jokes aside, Tony was a bit uncomfortable with her haste. He'd had a basic safety review done to make sure it wouldn't electrocute anyone, but that was about the extent of it. He wasn't even sure what sort of label to put on it. The lawyers weren't of much help when he asked them, so he looked up Food and Drug Administration regulations on his own and decided that a "for research use only" sticker was probably the most appropriate.

This was not a finished product and no one should be under the impression that it was, Tony thought.

| THREE |

Apple Envy

For a young entrepreneur building a business in the heart of Silicon Valley, it was hard to escape the shadow of Steve Jobs. By 2007, Apple's founder had cemented his legend in the technology world and in American society at large by bringing the computer maker back from the ashes with the iMac, the iPod, and the iTunes music store. In January of that year, he unveiled his latest and biggest stroke of genius, the iPhone, before a rapturous audience at the Macworld conference in San Francisco.

To anyone who spent time with Elizabeth, it was clear that she worshipped Jobs and Apple. She liked to call Theranos's blood-testing system "the iPod of health care" and predicted that, like Apple's ubiquitous products, it would someday be in every household in the country.

In the summer of 2007, she took her admiration for Apple a step further by recruiting several of its employees to Theranos. One of them was Ana Arriola, a product designer who'd worked on the iPhone.

Ana's first meeting with Elizabeth was at Coupa Café, a hip coffee and sandwich place in Palo Alto that had become her favorite haunt outside the office. After filling her in on her background and her travels to Asia, Elizabeth told Ana she envisioned building a disease map of each person through Theranos's blood tests. The company would then be able to reverse engineer illnesses like cancer with mathematical models that would crunch the blood data and predict the evolution of tumors.

It sounded impressive and world changing to a medical neophyte like Ana, and Elizabeth seemed brilliant. But given that Ana would be leaving behind fifteen thousand Apple shares if she joined Theranos, she wanted to get her wife Corrine's opinion. She arranged to meet Elizabeth again in Palo Alto, this time with Corrine present. Any hesitations she had were put to rest when Elizabeth made a big impression on Corrine too.

Ana joined Theranos as its chief design architect. This mostly meant she was responsible for the overall look and feel of the Edison. Elizabeth wanted a software touchscreen similar to the iPhone's and a sleek outer case for the machine. The case, she decreed, should have two colors separated by a diagonal cut, like the original iMac. But unlike that first iMac, it couldn't be translucent. It had to hide the robotic arm and the rest of the Edison's innards.

She'd contracted out the case's design to Yves Béhar, the Swiss-born industrial designer whose reputation in the Valley was second only to Apple's Jony Ive. Béhar came up with an elegant black-and-white design that proved difficult to build. Tony Nugent and Dave Nelson spent countless hours molding sheet metal in an attempt to get it right.

The case wouldn't conceal the loud noises the robotic arm made, but Ana was satisfied that it would at least make the device presentable when Elizabeth took it out on demos.

Ana felt that Elizabeth could use a makeover herself. The way she dressed was decidedly unfashionable. She wore wide gray pantsuits and Christmas sweaters that made her look like a frumpy accountant. People in her entourage like Channing Robertson and Don Lucas were beginning to compare her to Steve Jobs. If so, she should dress the part, she told her. Elizabeth took the suggestion to heart. From that point on, she came to work in a black turtleneck and black slacks most days.

Ana was soon joined at Theranos by Justin Maxwell and Mike Bauerly, two other recruits hired to work on the design of the Edison's software and other parts of the system that patients would interact with, like the packaging for the cartridges. Ana and Justin had worked together at Apple and knew Mike through his girlfriend, who had been a colleague of theirs there. It wasn't long before the Apple transplants began noticing that Elizabeth and Theranos had their quirks. Ana would arrive early every morning for a daily seven-thirty meeting with Elizabeth to update her on design issues. When she pulled her car into the parking lot, Ana would find her jamming to loud hip-hop music in her black Infiniti SUV, the blond streaks in her hair bouncing wildly.

One day, as Justin walked into her office to update her on a project, Elizabeth motioned him over excitedly, saying she wanted to show him something. She pointed to a nine-inch-long metal paperweight on her desk. Etched on it was the phrase, "What would you attempt to do if you knew you could not fail?" She'd positioned it so the words were facing her and clearly found it inspiring.

Having an idealistic boss wasn't a bad thing, but there were other aspects of working at Theranos that were less pleasant. One of them was having to do daily battle with Matt Bissel, the head of IT, and his sidekick, Nathan Lortz. Bissel and Lortz had the company's computer network set up in such a way that information was split into silos, hampering communication between employees and departments. You couldn't even exchange instant messages with a coworker. The chat ports were blocked. It was all in the name of protecting proprietary information and trade secrets, but the end result was hours of lost productivity.

The situation got so frustrating that Justin stayed up late one night and wrote a long email screed to Ana about it.

"We have lost sight of our business objective. Did this company set out to 'put a bunch of people in a room and prevent them from doing illegal things,' or did it set out to 'do something amazing with the best people, as quickly as possible'?" he fumed.

Justin and Mike also got the distinct impression that Bissel and Lortz were spying on them and reporting their findings back to Elizabeth. The IT team always wanted to know what programs they were running on their computers and at times turned suspiciously friendly in what felt like transparent attempts to elicit seditious gossip. The snooping wasn't confined to the IT guys. Elizabeth's administrative assistants would friend employees on Facebook and tell her what they were posting there.

One of the assistants kept track of when employees arrived and when they left so that Elizabeth knew exactly how many hours everyone put in. To entice people into working longer days, she had dinner catered every evening. The food often didn't arrive until eight or eight thirty, which meant that the earliest you got out of the office was ten.

The strange atmosphere got even stranger when the Theranos board convened once a quarter.

Employees were instructed to appear busy and not to make eye contact with the board members when they walked through the office. Elizabeth ushered them into a big glass conference room and pulled down the shades. It felt like CIA agents conducting secret debriefings with an undercover operative.

ONE EVENING, Ana gave Justin and Aaron Moore, one of the engineers, a ride back to San Francisco. Aaron had dropped out of a Ph.D. program in microfluidics at MIT and come to work at Theranos in September 2006 after spotting a small job ad in a trade publication. He'd worked at the company nearly a year by the time Ana and Justin came on board. Aaron was smart enough to have gone to college at Stanford and grad school at MIT, but he didn't take himself too seriously. He was originally from Portland, Oregon, and had the Portlandian hipster's look: shaggy hair, a three-day beard, and earrings. He was also witty, all of which made him the one person at Theranos the Apple transplants could relate to.

Ana, Justin, and Aaron all lived in San Francisco and commuted by car or train to the office. During their drive home that evening, Aaron shared some gripes he had with his new colleagues as they sat in traffic in Ana's Prius. In case they hadn't noticed yet, people were constantly getting fired at Theranos,

Aaron told them. Ana and Justin had definitely noticed. The Ed Ku layoffs had just taken place. In addition to Ed, twenty other people had lost their jobs. It happened so fast that Ed had left a bunch of work tools behind, including a nice set of X-Acto precision cutting knives that Justin had fished out of a wastebasket and claimed as his own.

Aaron mentioned that he was also troubled by the study with cancer patients in Tennessee. They'd never gotten the microfluidic system anywhere close to working properly and certainly not well enough to use on live patients, and yet Elizabeth had pushed ahead with the study. The shift to the new machine Tony built was an improvement, but Aaron felt they still didn't have a good read on its performance. The engineering and chemistry groups weren't communicating. Each was running tests on the parts of the system it was responsible for, but no one was conducting overall system tests.

Ana listened with rising unease. She'd assumed Theranos had perfected its blood-testing technology if it was going to be used on patients. Now Aaron was telling her it was still very much a work in progress. Ana knew the Tennessee study involved people dying of cancer. It bothered her to think they might be used as guinea pigs to test a faulty medical device.

What Ana and Aaron didn't know and what might have allayed their concerns somewhat is that the test results Theranos generated from the cancer patients' blood would not be used to make any changes to their treatments. They were to be used only for research purposes, to help Pfizer assess the effectiveness of Theranos's technology. But that was never clear to most Theranos employees because Elizabeth never explained the terms of the study.

The next morning, Ana reached out to the person who'd introduced her to Theranos: her former Apple colleague Avie Tevanian. Avie was on Theranos's board of directors. He was the one who'd put out feelers to Ana several months earlier and arranged for her to meet Elizabeth. Ana met Avie at a Peet's Coffee in Los Altos and mentioned what she'd learned from Aaron Moore. She worried that Theranos was crossing an ethical line with the Tennessee study. Avie listened intently and told Ana he was beginning to have doubts of his own about the company.

AVIE WAS ONE of Steve Jobs's oldest and closest friends. They'd worked together at NeXT, the software company Jobs created after being ousted from Apple in the mid-1980s. When Jobs returned to Apple in 1997 he'd brought Avie over with him and made him the head of software engineering. A grueling decade later, Avie had called it quits. He'd made more money than he knew what to do with and wanted to enjoy more time with his wife and two kids. A few months into his retirement, a headhunter recruiting new directors for Theranos had approached him.

Like Ana, Avie's first meeting with Elizabeth had been at Coupa Café. She'd come across as a bright young lady who was passionate about what she was doing, exactly the qualities you looked for in an entrepreneur. Her eyes had lit up when he volunteered some pieces of management wisdom he'd learned at Apple. His long association with Jobs seemed an object of fascination to her. After their encounter, Avie had agreed to join the Theranos board and bought $1.5 million of company stock in its late 2006 offering.

The first couple of board meetings Avie attended had been relatively uneventful, but, by the third one, he'd begun to notice a pattern. Elizabeth would present increasingly rosy revenue projections based on the deals she said Theranos was negotiating with pharmaceutical companies, but the revenues wouldn't materialize. It didn't help that Henry Mosley, the chief financial officer, had been fired soon after Avie became a director. At the last board meeting he'd attended, Avie had asked more pointed questions about the pharmaceutical deals and been told they were held up in legal review. When he'd asked to see the contracts, Elizabeth had said she didn't have any copies readily available.

There were also repeated delays with the product's rollout and the explanation for what needed to be fixed kept changing. Avie didn't pretend to understand the science of blood testing; his expertise was software. But if the Theranos system was in the final stages of fine-tuning as he'd been told, how could a completely different technical issue be the new holdup every quarter? That didn't sound to him like a product that was on the cusp of commercialization.

In late October 2007, he attended a meeting of the board's compensation committee. Don Lucas, the board's chairman, told the committee members that Elizabeth planned to create a foundation for tax-planning purposes and wanted the committee to approve a special grant of stock to it. Avie had noticed how much Don doted on Elizabeth. The old man treated her like a granddaughter. A portly gentleman with white hair who liked to wear broad-brim hats, Don was in his late seventies and was part of an older generation of venture capitalists who approached venture investing as if it were a private club. He'd mentored one famous entrepreneur in Larry Ellison. In Elizabeth, he clearly thought he'd found another.

Except Avie didn't think it was good corporate governance to do what Elizabeth wanted. Since she would control the foundation, she would also control the voting rights associated with the new stock, which would increase her overall voting stake. Avie didn't think it was in other shareholders' interest to give the founder more power. He objected.

Two weeks later, he received a call from Don

asking if they could meet. Avie drove to the old man's office on Sand Hill Road. Elizabeth was really upset, Don informed him when he got there. She felt he was behaving unpleasantly during board meetings and didn't think he should be on the board anymore. Don asked if he wanted to resign. Avie expressed surprise. He was just fulfilling his duties as a director; asking questions was one of them. Don agreed and said he thought Avie was doing an excellent job. Avie told Don he wanted to take a few days to think things over.

When he got back to his house in Palo Alto, he decided to go back and look at all the documents he'd been given over the previous year as a board member, including the investment materials he'd received before he bought his shares. As he read them over, he realized that everything about the company had changed in the space of a year, including Elizabeth's entire executive team. Don needed to see these, he thought.

IN THE MEANTIME, Ana Arriola was getting antsy. Ana was by nature excitable. She spoke quickly and was a constant whirlwind of activity. Most of the time, it was positive energy that she channeled into her work to great effect. But at times it could also turn into stress, anxiety, and drama.

After their coffee, she'd stayed in contact with Avie and had learned from her former Apple colleague that Elizabeth wanted him off the board. She didn't know what had prompted their rift, but it was an ominous development.

Ana's own relationship with Elizabeth was deteriorating. Elizabeth didn't like being told no, and Ana had done so on several occasions when she'd found a demand Elizabeth made unreasonable. She was also getting put off by her secrecy. A designer might not be as crucial to this little enterprise as an engineer or a chemist, but she still needed to be in the information loop about the product's development to do her job properly. Yet Elizabeth kept Ana on a need-to-know basis.

During one of their early morning meetings, Ana confronted Elizabeth with what she'd heard from Aaron Moore about problems with the Theranos system. If they were still working out kinks in the technology, wasn't it preferable to put the Tennessee study on pause and concentrate on fixing the problems first? They could always restart it once they got the machine working reliably, she told her.

Elizabeth flatly rejected the idea. Pfizer and every other big drugmaker wanted her blood-testing system and Theranos was going to be a great company, she said. If Ana wasn't happy, then perhaps she should reflect on whether this was the right place for her.

"Think about it and then tell me what you want to do," she said.

Ana went back to her desk and stewed for several

hours. She couldn't shake the thought that forging on with the Tennessee study wasn't the right thing to do. The fact that Elizabeth wanted Avie to leave the board was also unsettling. Ana trusted Avie and considered him a friend. If Avie and Elizabeth had a beef, she was inclined to side with Avie.

By midafternoon, Ana had made up her mind. She wrote up a brief resignation letter and printed out two copies, one for Elizabeth and one for HR. Elizabeth was out of the office by then, so she slipped the letter under her door. On her way out, she typed out a quick email to let her know where to find it.

Elizabeth emailed her back thirty minutes later, asking her to please call her on her cell phone. Ana ignored her request. She was done with Theranos.

DON LUCAS DIDN'T USE EMAIL. He'd seen his share of litigation over the years, including a wave of class-action lawsuits targeting Oracle in the early 1990s, and didn't like the idea of leaving behind an electronic paper trail that might one day be used against him in court. If Avie wanted Don to see what he'd found, he'd have to show it to him in person. He reached out to Don's two assistants and set up another meeting.

On the appointed day, Avie showed up at Don's office with hard copies of all the documents he had been given as a Theranos director. It amounted to hundreds of pages. Taken together, they betrayed a series of irreconcilable discrepancies, he told Don. The board had a problem on its hands, he said. It was possible Theranos could be fixed, but it wasn't going to happen the way Elizabeth was managing things. He suggested they bring in some adult supervision.

"Well, I think you should resign," Don replied. He quickly added, "What are you planning to do with that stack of papers?"

Avie was taken aback. Don didn't even seem interested in hearing him out. The older man seemed concerned only with whether he was going to escalate the matter to the full board. After turning the situation over in his mind for a few moments, Avie decided to stand down. He'd retired from Apple for a reason. He didn't need the aggravation.

"OK, I'll resign and I'll leave these papers with you," he said.

As Avie got up to leave, Don said there was something else they needed to discuss. Shaunak Roy, Theranos's first employee and de facto cofounder, was leaving the company and selling most of his founder's shares back to Elizabeth. She needed the board to waive the company's rights to repurchase the stock. Avie didn't think that was a good idea but told Don to have the board vote the motion without him since he was resigning.

"One more thing, Avie," Don said. "I need you to waive your own rights to buy the shares."

Avie was starting to get ticked off. He was being asked to put up with a lot. He told Don to have Michael Esquivel, Theranos's general counsel, send over the requisite documents. He would review them but made no promises.

When the documents arrived, Avie read them carefully and concluded that, once the company itself waived its rights to repurchase Shaunak's shares, it was entirely within his and other shareholders' rights to buy some of them. He also noticed that Elizabeth had negotiated a sweetheart deal: Shaunak was willing to part with his 1.13 million shares for $565,000. That translated to 50 cents a share, an 82 percent discount to what he and other investors had paid more than a year earlier in Theranos's last funding round. Some discount was warranted because Avie's shares were preferred shares with higher claims on the company's assets and earnings while Shaunak's shares were common ones, but a discount that big was unheard of.

Avie decided to exercise his rights and told Esquivel he wanted to acquire the pro-rata portion of Shaunak's stock he was entitled to. The request did not go down well. A tense email exchange ensued between the two men that stretched into the Christmas holiday.

At 11:17 p.m. on Christmas Eve, Esquivel sent Avie an email accusing him of acting in "bad faith" and warned him that Theranos was giving serious consideration to suing him for breach of his fiduciary duties as a board member and for public disparagement of the company.

Avie was astonished. Not only had he done no such things, in all his years in Silicon Valley he had never come close to being threatened with a lawsuit. All over the Valley, he was known as a nice guy. A teddy bear. He didn't have any enemies. What was going on here? He tried getting in touch with other members of the board, but none would respond to his calls.

Unsure what to do, Avie consulted a friend who was a lawyer. Thanks to his Apple wealth, his personal balance sheet was bigger than Theranos's, so the prospect of costly litigation didn't really scare him. But after he filled his friend in on everything that had happened, the friend asked a question that helped him put the situation in perspective: "Given everything you now know about this company, do you really want to own more of it?"

When Avie thought about it, the answer was no. Besides, it was the season of giving and rejoicing. He decided to let the matter rest and to put Theranos behind him. But before doing so, he wrote a parting letter to Don and emailed it to his assistants, along with a copy of the waiver the company had pressured him to sign.

The brutal tactics used to get him to sign the waiver, he wrote, had confirmed "some of the worse concerns" he'd raised with Don about the way the company was being run. He didn't blame

Michael Esquivel, he added, because it was clear the attorney was just acting on orders from above. He closed the letter with

> I do hope you will fully inform the rest of the Board as to what happened here. They deserve to know that by not going along 100% "with the program" they risk retribution from the Company/Elizabeth.
>
> . . .
>
> Warmly,
> Avie Tevanian

| FOUR |

Goodbye East Paly

In early 2008, Theranos moved to a new building on Hillview Avenue in Palo Alto. It was the Silicon Valley equivalent of moving from the South Bronx to Midtown Manhattan.

Appearances in the Valley are paramount and for three years Theranos had been operating on the wrong side of the tracks. The "tracks" in this case were Route 101, otherwise known as the Bayshore Freeway. It separates Palo Alto, one of the most affluent towns in America, from its poorer sibling East Palo Alto, which once held the dubious distinction of being the country's murder capital.

The company's old office was on the East Palo Alto side of the four-lane highway, next to a machine shop and across the street from a roofing contractor. It wasn't the type of neighborhood wealthy venture capitalists liked to be seen in. The new address, by contrast, was right next to the Stanford campus and around the corner from Hewlett-Packard's plush headquarters. It was pricey real estate that signaled Theranos was graduating to the big leagues.

Don Lucas was pleased with the move. During a conversation with Tony Nugent, he made clear his disdain for the old location. "It's nice to finally get Elizabeth out of East Paly," he told Tony.

The move was not fun for the person who had to make it happen, however. That job fell to Matt Bissel, the head of IT. Bissel was one of Elizabeth's most trusted lieutenants. He'd joined Theranos in 2005 as employee number 17 and took his duties seriously. In addition to being responsible for the company's IT infrastructure, his role also included security. He was the one who'd done the forensic analysis of the computer evidence for the Michael O'Connell lawsuit.

Planning the move had taken up a big chunk of Matt's time over the past several months. On Thursday, January 31, 2008, everything finally seemed ready. The movers were scheduled to arrive first thing the next morning to haul everything away.

But at four that afternoon, Matt got pulled into a conference room with Michael Esquivel and Gary Frenzel. Elizabeth was conferenced in by phone from Switzerland, where she was conducting a second demonstration for Novartis some fourteen months after the faked one that had led to Henry Mosley's departure. She'd just learned that the landlord would charge them rent for the month of February if they didn't clear the premises by midnight. There was no way she was going to let that happen, she said.

She instructed Matt to call the moving company and have the movers come immediately. Matt thought the odds that would happen were very low but agreed to try. He stepped out of the conference room and made the call. The moving company's dispatcher laughed at him. No sir, rescheduling a corporate move at the eleventh hour wasn't possible, he was told.

Elizabeth was undeterred. She told Matt to call another moving company she had once used and to give them the job. Unlike the first company, this one wasn't unionized. She was sure it would be more flexible. But when Matt called the second moving company and explained the situation, a person there strongly advised him to drop the idea. Unionized moving companies were all mob controlled, the person said. What Theranos was proposing to do risked devolving into violence.

Even after hearing that sobering answer, Elizabeth wouldn't let it go. Matt and Gary tried to reason with her by citing other obstacles. Gary raised the issue of their stockpile of blood samples.

Supposing they managed to get a crew to come that day; the movers wouldn't unload everything at the new address until tomorrow, he pointed out. How would they keep the blood at the proper temperature in the meantime? Elizabeth said they could use refrigerated trucks and keep them running in the parking lot overnight.

After several crazed hours, Matt was finally able to talk some sense into her by pointing out that even if they somehow cleared the building by 11:59 p.m. that night, they would still have to conduct walkthroughs with state officials to demonstrate that they had properly disposed of any hazardous materials. Theranos was a biotech company, after all. Those walkthroughs would take weeks to schedule and no new tenant would be able to move in until they had occurred.

In the end, the move took place the next day as originally planned, but the episode was the final straw for Matt. Part of him admired Elizabeth. She was one of the smartest people he'd ever met and she could be a really inspiring and energizing leader. He often joked that she could sell ice cream to Eskimos. But another part of him was tiring of her unpredictability and the constant chaos at the company.

One aspect of Matt's job had become increasingly distasteful to him. Elizabeth demanded absolute loyalty from her employees and if she sensed that she no longer had it from someone, she could turn on them in a flash. In Matt's two and a half years at Theranos, he had seen her fire some thirty people, not counting the twenty or so employees who lost their jobs at the same time as Ed Ku when the microfluidic platform was abandoned.

Every time Elizabeth fired someone, Matt had to assist with terminating the employee. Sometimes, that meant more than just revoking the departing employee's access to the corporate network and escorting him or her out of the building. In some instances, she asked him to build a dossier on the person that she could use for leverage.

There was one case in particular that Matt regretted helping her with: that of Henry Mosley, the former chief financial officer. After Elizabeth fired Mosley, Matt had stumbled across inappropriate sexual material on his work laptop as he was transferring its files to a central server for safekeeping. When Elizabeth found out about it, she used it to claim it was the cause of Mosley's termination and to deny him stock options.

Matt had reported to Mosley until he left and thought he'd done an excellent job of helping Elizabeth raise money for Theranos. He clearly shouldn't have browsed porn on a work-issued laptop, but Matt didn't think it was a capital offense that merited blackmailing him. And besides, it had been found **after** the fact. Saying it was the reason Mosley was fired simply wasn't true.

The way John Howard was treated also bothered

him. When Matt reviewed all the evidence assembled for the Michael O'Connell lawsuit, he didn't see anything proving that Howard had done anything wrong. He'd had contact with O'Connell but he'd declined to join his company. Yet Elizabeth insisted on connecting the dots a certain way and suing him too, even though Howard had been one of the first people to help her when she dropped out of Stanford, letting her use the basement of his house in Saratoga for experiments in the company's early days. (Theranos later dropped the case against its three ex-employees when O'Connell agreed to sign his patent over to the company.)

Matt had long wanted to start his own IT consulting firm and he decided this was the time to walk away and do it. When he informed Elizabeth of his decision, she looked at him in utter disbelief. She couldn't comprehend how he could possibly trade in a job at a company that was going to revolutionize health care and change the world for that. She tried to entice him to stay with a raise and a promotion, but he turned her down.

During his last couple of weeks at Theranos, what Matt had seen happen to numerous other employees started happening to him. Elizabeth wouldn't speak to him anymore or even look at him. She offered one of his IT colleagues, Ed Ruiz, his position if Ed agreed to dig through Matt's files and emails. But Ed was good friends with Matt and refused. In any case, there was nothing to find. Matt was squeaky-clean. Unlike Henry Mosley, he was able to keep his stock options and to exercise them. He left Theranos in February 2008 and started his own firm. Ed Ruiz joined him a few months later.

THERANOS'S NEW OFFICE in Palo Alto was nice, but it was actually too big for a startup that had just shrunk back down to fifty people after the Ed Ku layoffs. The main floor was a long rectangular expanse. Elizabeth insisted on clustering employees on one side of it, leaving a big empty stretch of space on the other. Once or twice, Aaron Moore tried to put it to use by coaxing several colleagues into a game of indoor soccer.

Aaron grew closer to Justin Maxwell and Mike Bauerly after Ana Arriola's sudden departure. Ana hadn't given any of them a heads-up that she planned on quitting. She'd just marched out one day and hadn't come back. It unsettled Justin the most because Ana was the one who'd talked him into leaving Apple to come to Theranos, but he tried to maintain a positive attitude. He told himself that if the company was moving to prime Palo Alto office space, then it must be doing something right.

Shortly after the move, Aaron and Mike decided to conduct some informal "human factors" research with two of the Edison prototypes Tony Nugent and Dave Nelson had built. It was engineering-speak for putting them in people's hands and seeing how they interacted with them. Aaron was curious to know how people handled pricking their fingers and the subsequent steps required to get the blood into the cartridge. He'd pricked his own finger so much while running internal tests that he no longer had any feeling in it.

With Tony's permission, they put the Edisons in the trunk of Aaron's Mazda and drove up to San Francisco. Their plan was to take them around to friends' startups in the city. First, they stopped at Aaron's apartment in San Francisco's Mission District to do some staging. They placed the machines on the wooden coffee table in Aaron's living room and made sure they had everything else they needed: the cartridges, the lancets to draw the blood, and the small syringes called "transfer pens" used to put the blood in the cartridge.

Aaron took photos with his digital camera to document what they were doing. The Yves Béhar cases weren't ready yet, so the devices had a primitive look. Their temporary cases were made from gray aluminum plates bolted together. The front plate tilted upward like a cat door to let the cartridge in. A rudimentary software interface sat atop the cat door at an angle. Inside, the robotic arm made loud, grinding sounds. Sometimes, it would crash against the cartridge and the pipette tips would snap off. The overall impression was that of an eighth-grade science project.

When Aaron and Mike arrived at their friends'

offices, they were greeted with chuckles and cups of coffee. Everyone was a good sport, though, and agreed to go along with their little experiment. One of the stops was at Bebo, a social networking startup that was acquired by AOL a few weeks later for $850 million.

As the day progressed, it became apparent that one pinprick often wasn't enough to get the job done. Transferring the blood to the cartridge wasn't the easiest of procedures. The person had to swab his finger with alcohol, prick it with the lancet, apply the transfer pen to the blood that bubbled up from the finger to aspirate it, and then press on the transfer pen's plunger to expel the blood into the cartridge. Few people got it right on their first try. Aaron and Mike kept having to ask their test subjects to prick themselves multiple times. It got messy. There was blood everywhere.

These difficulties confirmed what Aaron already suspected: the company was underestimating this part of the process. To assume that a fifty-five-year-old patient in his or her home was going to immediately master it was wishful thinking. And if you didn't get this part right, it didn't matter how well the rest of the system functioned; you weren't going to get good results. When they got back to the office, Aaron passed on his findings to Tony and Elizabeth, but he could tell they didn't think they were a priority.

Aaron was getting frustrated and disillusioned.

He'd initially bought into Elizabeth's vision and found work at Theranos exciting. But after nearly two years, he was getting burned out. Among other issues, he didn't get along with Tony, who'd become his boss. To get out from under him, he had asked to transfer from engineering to sales. He'd even spent a recent Saturday driving around shopping for a suit in the hope that Elizabeth would let him tag along on her trip to Switzerland. She didn't, but she seemed to at least be taking his transfer request under advisement.

A few days after the San Francisco excursion, Aaron was sipping a beer at home and downloading the pictures he'd taken when an idea for a joke came to him. Using Photoshop software, he took one of the pictures—it showed the twin Edison prototypes sitting side by side on dinner mats on his coffee table—and made a fake Craigslist ad. Above the photo and under a headline that read, "Theranos Edison 1.0 'readers'—mostly functional—$10,000 OBO," he wrote:

> Up for grabs is a rare matching set of Theranos point-of-care diagnostic "Edison" devices. Billed as the "iPod of healthcare," the Edison is a semi-portable immunochemistry platform capable of performing multiplexed protein assays on a finger-stick sample of human or animal whole blood . . .
>
> I bought these units recently when I thought I was at risk of succumbing to septic shock. Now

> that I've tested my protein C and realized that I'm safely in the 4 ug/mL range, I no longer need a pre-production blood analytic device. My loss is your gain!
>
> $10k for the pair, $6000 apiece, OBO—would also be willing to consider trade for a comparable pre-clinical diagnostic device (i.e., Roche, Becton-Coulter [**sic**], Abaxis, Biosite, etc.). Comes with a supply of single use cartridges, pelican shipping cases, AC adapter, EU power adapters, and assorted blood collection accessories, leeches, etc.

Aaron printed out the mock ad and took it with him to work the next day. When Justin and Mike spotted it on his desk, they thought it was hilarious. Mike decided it deserved a bigger audience and posted it on the wall in the men's room.

Then all hell broke loose. Someone took the ad down and brought it to Elizabeth, who thought it was real. She convened an emergency meeting of the senior managers and the lawyers. She was treating it as a full-blown case of industrial espionage and wanted an immediate investigation to find the culprit.

Aaron decided he better fess up before things got further out of control. He sheepishly came forward and confessed to Tony. It was meant as an innocent prank, he explained. He thought people would find it amusing. Tony seemed understanding. He'd taken part in a few pranks of his own at Logitech when he worked there. But he warned Aaron that Elizabeth was furious.

Later in the day, she called Aaron into her office and stared at him with dagger eyes. She was deeply disappointed in him, she told him. She didn't find his little stunt funny at all, and neither did other employees. It was disrespectful to the people who'd worked so hard to make the product. He could forget about joining the sales team. She couldn't put him in front of customers. This showed he represented the company poorly. Aaron went back to his cubicle with the knowledge that he was now squarely in Elizabeth's doghouse.

A MOVE TO SALES would probably have been ill-advised anyway. Unbeknownst to Aaron, trouble was brewing in that corner of the company. A new employee named Todd Surdey had come on board to head up sales and marketing, a role previously played by Elizabeth herself.

Todd was the consummate sales executive. Before joining Theranos, he had worked at several established companies, including most recently the German enterprise software juggernaut SAP. He was fit and good-looking, wore nice suits, and rolled up every day in a fancy BMW. During lunchtime, he pulled a carbon fiber road bike out of his trunk and went on rides in the nearby hills. Aaron liked to cycle too and accompanied Todd a few times in an attempt to buddy up to him before his prank put him in the penalty box with Elizabeth.

Todd's two sales subordinates were based on the East Coast, where all the big pharmaceutical companies were headquartered. One of them was Susan DiGiaimo, an employee who operated out of her home in New Jersey and had worked for Theranos for nearly two years. Susan had accompanied Elizabeth on numerous sales pitches to drugmakers and had listened uncomfortably as she promised them the moon. When the drugmakers' executives asked if the Theranos system could be customized to suit their needs, Elizabeth would always answer, "Absolutely."

Soon after he started, Todd began asking Susan a lot of questions about the revenues Elizabeth was projecting from her deals with the drugmakers. She kept a spreadsheet with detailed revenue forecasts. The numbers were big, in the tens of millions of dollars for each deal. Susan told Todd that, based on what she knew, they were vastly overinflated.

Moreover, no significant revenues would materialize unless Theranos proved to each partner that its blood system worked. To that effect, each deal provided for an initial tryout, a so-called validation phase. Some companies, like the British drugmaker AstraZeneca, weren't willing to pay more than $100,000 for the validation phase, and all could walk away if they weren't happy with the results.

The 2007 study in Tennessee was the validation

phase of the Pfizer contract. Its objective was to prove that Theranos could help Pfizer gauge cancer patients' response to drugs by measuring the blood concentration of three proteins the body produces in excess when tumors grow. If Theranos failed to establish any correlation between the patients' protein levels and the drugs, Pfizer could end their partnership and any revenue forecast Elizabeth had extrapolated from the deal would turn out to be fiction.

Susan also shared with Todd that she had never seen any validation data. And when she went on demonstrations with Elizabeth, the devices often malfunctioned. A case in point was the one they'd just conducted at Novartis. After the first Novartis demo in late 2006 during which Tim Kemp had beamed a fabricated result from California to Switzerland, Elizabeth had continued to court the drugmaker and had arranged a second visit to its headquarters in January 2008.

The night before that second meeting, Susan and Elizabeth had pricked their fingers for two hours in a hotel in Zurich to try to establish some consistency in the test results they were getting, to no avail. When they showed up at Novartis's Basel offices the next morning, it got worse: all three Edison readers produced error messages in front of a room full of Swiss executives. Susan was mortified, but Elizabeth kept her composure and blamed a minor technical glitch.

Based on the intel he was getting from Susan and from other employees in Palo Alto, Todd became convinced that Theranos's board was being misled about the company's finances and the state of its technology. He took his concerns to Michael Esquivel, the general counsel with whom he had established good rapport.

Michael, it turned out, was developing his own suspicions. During a lunchtime run with a colleague from the new office to the Stanford Dish and back, he mentioned not feeling too good about Theranos's pharmaceutical partnerships. He wouldn't say more, but the colleague could tell something was bothering him.

In March 2008, Todd and Michael approached Tom Brodeen, one of Theranos's board members, and told him the revenue projections Elizabeth was touting to the board weren't grounded in reality. They were hugely exaggerated and impossible to reconcile with the unfinished state of the product, they said.

Brodeen was a seasoned businessman in his mid-sixties who had headed one of the big consulting firms as well as several technology companies. He hadn't been on the Theranos board long, having joined at the request of Don Lucas in the fall of 2007. Given how new he was as a director, he advised Todd and Michael to take their account directly to Lucas, the board's chairman.

Coming just months after Avie Tevanian had

raised similar concerns, Lucas took the matter seriously this time. In a way, he couldn't afford not to: Todd was the son-in-law of one of Theranos's investors, the venture capitalist B. J. Cassin. Cassin and Lucas were longtime friends. They had both invested in Theranos at the same time, during the startup's Series B round in early 2006.

Lucas convened an emergency meeting of the board in his office on Sand Hill Road. Elizabeth was asked to wait outside the door while the other directors—Lucas, Brodeen, Channing Robertson, and Peter Thomas, the founder of an early stage venture capital firm called ATA Ventures—conferred inside.

After some discussion, the four men reached a consensus: they would remove Elizabeth as CEO. She had proven herself too young and inexperienced for the job. Tom Brodeen would step in to lead the company for a temporary period until a more permanent replacement could be found. They called in Elizabeth to confront her with what they had learned and inform her of their decision.

But then something extraordinary happened.

Over the course of the next two hours, Elizabeth convinced them to change their minds. She told them she recognized there were issues with her management and promised to change. She would be more transparent and responsive going forward. It wouldn't happen again.

Brodeen wasn't exactly dying to come out of

retirement to run a startup in a field in which he had no expertise, so he took a neutral stance and watched as Elizabeth used just the right mix of contrition and charm to gradually win back his three board colleagues. It was an impressive performance, he thought. A much older and more experienced CEO skilled in the art of corporate infighting would have been hard-pressed to turn the situation around like she had. He was reminded of an old saying: "When you strike at the king, you must kill him." Todd Surdey and Michael Esquivel had struck at the king, or rather the queen. But she'd survived.

THE QUEEN DIDN'T WASTE any time putting down the rebellion. Elizabeth fired Surdey first and Esquivel a few weeks later.

To Aaron Moore, Mike Bauerly, and Justin Maxwell, this new purge was one more negative development. They weren't privy to what had happened, but they did know that Theranos had lost two good employees. Todd and Michael weren't just nice guys they got along with, they were smart and principled colleagues. In Mike Bauerly's words, they were cut from good cloth.

The firings caused Justin to further sour on Theranos. The staff turnover was like nothing he'd ever experienced before and he was troubled by what he saw as a culture of dishonesty at the company.

The worst offender was Tim Kemp, the head of the software team. Tim was a yes-man who never leveled with Elizabeth about what was feasible and what wasn't. For instance, he'd contradicted Justin and assured her they could write the Edison software's user interface faster in Flash than in JavaScript. The very next morning, Justin had spotted a Learn Flash book on his desk.

Elizabeth never reprimanded Tim, even when obvious examples of his duplicity were brought to her attention. She valued his loyalty and, in her eyes, the fact that he never said no to her reflected a can-do attitude. It mattered little that many of his colleagues thought Tim was a mediocrity and a terrible manager.

There was one incident involving Elizabeth herself that also didn't sit well with Justin. During an email exchange one evening, he asked her for a piece of information he needed to write a section of software. She responded that she'd look for it when she was back at work the next morning. The clear implication was that she had gone home. But minutes later, he stumbled on her in Tony Nugent's office down the hall. Justin got angry and stormed off.

Elizabeth came by his office a little later to say she understood why he was upset, but she warned him, "Don't ever walk off on me again."

Justin tried to remind himself that Elizabeth was

very young and still had a lot to learn about running a company. In one of their last email exchanges, he recommended two management self-help books to her, **The No Asshole Rule: Building a Civilized Workplace and Surviving One That Isn't** and **Beyond Bullsh*t: Straight-Talk at Work,** and included their links on Amazon.com.

He quit two days later. His resignation email read in part:

> good luck and please do read those books, watch The Office, and believe in the people who disagree with you . . . Lying is a disgusting habit, and it flows through the conversations here like it's our own currency. The cultural disease here is what we should be curing before we try to tackle obesity . . . I mean no ill will towards you, since you believe in what I was doing and hoped I would succeed at Theranos. I feel like I owe you this bad attempt at an exit interview since we have no HR to officially record it.

Upset, Elizabeth called him into her office, told him she disagreed with his criticism, and asked him to resign "with dignity." Justin agreed to help smooth the transition by sending his colleagues an email with detailed instructions about where to find the various projects he'd been working on. But as he sat down to write it, he couldn't resist including a few personal thoughts on the state of those projects, earning him one last reprimand from Elizabeth.

Aaron Moore and Mike Bauerly stayed at Theranos a few more months, but their hearts were no longer in it. One of the nice features of the new office was that it had a big terrace above the building's entrance. Mike had furnished it with deck chairs and a hammock. Aaron and Mike would retreat there for long coffee breaks, the afternoon sun pleasantly warming their faces as they bantered.

Aaron felt someone needed to tell Elizabeth to pump the brakes and to stop pushing to commercialize a product that they were still trying to get to work. But for her to listen, the message had to come from one of the three senior managers—Tim, Gary, or Tony—and none of them were willing to tell her. Tony, who was under a lot of pressure from Elizabeth, finally had enough of hearing Aaron's complaints and asked him to leave the company. "Go find a place where you can be a big fish in a small pond," he told him.

Aaron agreed that it was time for him to go. To his surprise, Elizabeth tried to convince him to stay. It turned out she thought highly of him despite his prank. But his mind was made up. He resigned in June 2008. Mike Bauerly followed in December. Every member of the Apple contingent had now moved on, marking the end of a chaotic period for the company. Elizabeth had survived an aborted board coup and was back firmly in control. Remaining Theranos employees looked forward to calmer and quieter times. But their hopes would soon be dashed.

| FIVE |

The Childhood Neighbor

While Elizabeth was busy building Theranos, an old family acquaintance was taking an interest in what she was doing from afar. His name was Richard Fuisz. He was an entrepreneur–cum–medical inventor with a big ego and a colorful background.

The Holmes and Fuisz families had known each other for two decades. They first met in the 1980s as neighbors in Foxhall Crescent, a leafy neighborhood of stately homes in Washington, D.C., surrounded by woodlands and abutting the Potomac River.

Elizabeth's mother, Noel, and Richard's wife, Lorraine, struck up a close friendship. Both were stay-at-home mothers back then, raising children of similar ages. Lorraine's son was in Elizabeth's class at St. Patrick's Episcopal Day School, the neighborhood's private elementary school.

Noel and Lorraine were in and out of each other's houses. They shared a weakness for Chinese food and often went out for lunch while the children were in school. Elizabeth and her brother attended the Fuisz children's birthday parties and frolicked in the Fuiszes' pool. One evening, the power went out in the Fuisz home while Richard was away, so the Holmeses took Lorraine and her two children, Justin and Jessica, in for the night.

The relationship between their husbands wasn't as warm. While Chris Holmes had to make do on a government salary, Richard Fuisz was a successful businessman and wasn't shy about flaunting it. A licensed doctor, he had sold a company that made medical-training films for more than $50 million a few years earlier and drove a Porsche and a Ferrari. He was also a medical inventor who licensed out his patents and reaped the royalties. During one excursion the families made together to the zoo, Justin Fuisz remembers, Elizabeth's younger brother, Christian, told him, "My dad thinks your dad is an asshole." When Justin later repeated the comment to his mother, Lorraine chalked it up to jealousy.

Money was indeed a sore point in the Holmes household. Chris's grandfather, Christian Holmes II, had depleted his share of the Fleischmann fortune by living a lavish and hedonistic lifestyle on an island in Hawaii, and Chris's father, Christian III, had frittered away what was left during an unsuccessful career in the oil business.

Whatever simmering resentments Chris Holmes harbored did not prevent Noel Holmes and Lorraine Fuisz from being good friends. The two women stayed in regular contact even after the Holmeses moved away, first to California and then to Texas. When the Holmeses returned to Washington for a brief period in between, the Fuiszes took them out to a nice restaurant to celebrate Noel's fortieth birthday. Lorraine arranged the outing to make up for the fact that Chris hadn't thrown his wife a party.

Lorraine later visited Noel in Texas several times, and they also traveled to New York City together to shop and sightsee. They brought the children along once and booked rooms at the Regency Hotel on Park Avenue. In a photo from that trip, Elizabeth can be seen standing arm in arm between her mother and Lorraine in front of the hotel. She's wearing a light blue summer dress and pink bows in her hair. On subsequent trips, Noel and Lorraine left the children at home and stayed in an apartment the Fuiszes purchased in the Trump International Hotel and Tower on Central Park West.

In 2001, Chris Holmes hit a rough patch in his career. He had left Tenneco to take a position at Enron, Houston's most prominent corporation. When Enron's fraudulent practices were exposed and it went bankrupt in December of that year, he lost his job like thousands of other employees. In the aftermath, he paid a visit to Richard Fuisz in search of job leads and business advice. With one of his sons from a previous marriage, Fuisz had started a new company around one of his inventions: a thin strip that dissolved in the mouth and delivered drugs to the bloodstream faster than traditional pills. He and his son, Joe, ran it from a suite of offices in Great Falls, Virginia.

Chris Holmes came in looking haggard and glum, Joe Fuisz recalls. He mused aloud about trying his hand at consulting and indicated that he and Noel were desperate to move back to Washington. Having just purchased a new house in the affluent Beltway suburb of McLean, Richard Fuisz offered him use of the one he and Lorraine had just vacated across the street, rent-free. They hadn't bothered to list it yet. Chris mouthed a "thank you" but didn't take him up on the offer.

CHRIS AND NOEL HOLMES DID eventually move back to Washington four years later when Chris got a job at the World Wildlife Fund. At first, they stayed with friends in Great Falls while they looked for a new place to live. As Noel toured houses, she called Lorraine frequently to update her on her search.

Over lunch one day, the topic turned to Elizabeth

and what she was up to. Noel proudly told Lorraine that her daughter had invented a wrist device that could analyze a person's blood and started a company to commercialize it. The reality was that Theranos was already moving on from Elizabeth's original patch idea at that point, but that lost nuance hardly mattered in the chain of events Noel's lunchtime confidence unleashed.

When she got home, Lorraine repeated what Noel had told her to her husband, thinking it might be of interest to him as a fellow medical inventor. What she probably didn't anticipate is how he would react.

Richard Fuisz was a vain and prideful man. The thought that the daughter of longtime friends and former neighbors would launch a company in his area of expertise and that they wouldn't ask for his help or even consult him deeply offended him. As he would put it years later in an email, "The fact that the Holmes family was so willing to partake of our hospitality (New York apartment, dinners, etc.) made it particularly bitter to me that they would not ask for advice. Essentially the message was, 'I'll drink your wine but I won't ask you for advice in the very field that paid for the wine.' "

FUISZ HAD A HISTORY of taking slights personally and bearing grudges. The lengths he was willing to go to get even with people he perceived to have crossed him is best illustrated by his long and protracted feud with Vernon Loucks, the CEO of hospital supplies maker Baxter International.

Throughout the 1970s and early 1980s, Fuisz traveled a lot to the Middle East, which had become the biggest market for Medcom, his medical film business. On his way back, he usually spent a night in Paris or London and from there took the Concorde, the supersonic passenger jet operated by British Airways and Air France, back to New York. During one of these stopovers in 1982, he ran into Loucks at the Plaza Athénée hotel in Paris. At the time, Baxter was eager to expand into the Middle East. Over dinner, Loucks offered to buy Medcom for $53 million and Fuisz accepted.

Fuisz was supposed to stay on to head the new Baxter subsidiary for three years, but Loucks dismissed him shortly after the acquisition closed. Fuisz sued Baxter for wrongful termination, alleging that Loucks had fired him for refusing to pay a $2.2 million bribe to a Saudi firm to get Baxter off an Arab blacklist of companies that did business with Israel.

The two sides reached a settlement in 1986, under which Baxter agreed to pay Fuisz $800,000. That wasn't the end of it, however. When Fuisz flew to Baxter's Deerfield, Illinois, headquarters to sign the settlement, Loucks refused to shake his hand, angering Fuisz and putting him back on the warpath.

In 1989, Baxter was taken off the Arab boycott list, giving Fuisz an opening to seek his revenge.

He was leading a double life as an undercover CIA agent by then, having volunteered his services to the agency a few years earlier after coming across one of its ads in the classified pages of the **Washington Post**.

Fuisz's work for the CIA involved setting up dummy corporations throughout the Middle East that employed agency assets, giving them a non-embassy cover to operate outside the scrutiny of local intelligence services. One of the companies supplied oil-rig operators to the national oil company of Syria, where he was particularly well connected.

Fuisz suspected Baxter had gotten itself back in Arab countries' good graces through chicanery and set out to prove it using his Syrian connections. He sent a female operative he'd recruited to obtain a memorandum kept on file in the offices of the Arab League committee in Damascus that was in charge of enforcing the boycott. It showed that Baxter had provided the committee detailed documentation about its recent sale of an Israeli plant and promised it wouldn't make new investments in Israel or sell the country new technologies. This put Baxter in violation of a U.S. anti-boycott law, enacted in 1977, that forbade American companies from participating in any foreign boycott or supplying blacklist officials any information that demonstrated cooperation with the boycott.

Fuisz sent one copy of the explosive memo to

Baxter's board of directors and another to the **Wall Street Journal,** which published a front-page story about it. Fuisz didn't let the matter rest there. He subsequently obtained and leaked letters Baxter's general counsel had written to a general in the Syrian army that corroborated the memo.

The revelations led the Justice Department to open an investigation. In March 1993, Baxter was forced to plead guilty to a felony charge of violating the anti-boycott law and to pay $6.6 million in civil and criminal fines. The company was suspended from new federal contracts for four months and barred from doing business in Syria and Saudi Arabia for two years. The reputational damage also cost it a $50 million contract with a big hospital group.

For most people, this would have been ample vindication. But not for Fuisz. It irked him that Loucks had survived the scandal and remained CEO of Baxter. So he decided to subject his foe to one last indignity.

Loucks was a Yale alumnus and served as a trustee of the Yale Corporation, the university's governing body. He was also chairman of its fund-raising campaign. As he did every year in his capacity as a trustee, he was scheduled to attend Yale's commencement exercises in New Haven, Connecticut, that May.

Through his son Joe, who had graduated from Yale the year before, Fuisz got in touch with a student named Ben Gordon, who was the president of the Yale Friends of Israel association. Together, they organized a graduation day protest featuring "Loucks Is Bad for Yale" signs and leaflets. The crowning flourish was a turboprop plane Fuisz hired to fly over the campus trailing a banner that read, "Resign Loucks."

Three months later, Loucks stepped down as a Yale trustee.

DRAWING TOO CLOSE a parallel between Fuisz's vendetta against Loucks and the actions he would take with respect to Theranos would be an oversimplification, however.

As much as he was annoyed by what he perceived as the Holmeses' ingratitude, Fuisz was also an opportunist. He made his money patenting inventions he anticipated other companies would someday want. One of his most lucrative plays involved repurposing a cotton candy spinner to turn drugs into fast-dissolving capsules. The idea came to him when he took his daughter to a country fair in Pennsylvania in the early 1990s. He later sold the public corporation he formed to house the technology to a Canadian pharmaceutical company for $154 million and personally pocketed $30 million from the deal.

After Lorraine relayed what Noel had told her, Fuisz sat down at his computer in the sprawling seven-bedroom home they occupied in McLean and googled "Theranos." The house was so spacious that he had turned its great room, which had a high vaulted ceiling and a massive stone fireplace, into his personal study. His Jack Russell liked to lie in front of the fireplace while he worked.

Fuisz came upon the startup's website. The home page gave a cursory description of the microfluidic system Theranos was developing. Under the website's News tab, he also found a link to a radio interview Elizabeth had given to NPR's "BioTech Nation" segment a few months earlier, in May 2005. In the interview, she'd described her blood-testing system in more detail and explained the use she foresaw for it: at-home monitoring of adverse reactions to drugs.

Fuisz listened to the NPR interview several times while gazing out the window at the koi pond in his yard and decided there was some merit to Elizabeth's vision. But as a trained physician, he also spotted a potential weakness he could exploit. If patients were going to test their blood at home with the Theranos device to monitor how they were tolerating the drugs they were taking, there needed to be a built-in mechanism that would alert their doctors when the results came back abnormal.

He saw a chance to patent that missing element, figuring there was money to be made down the road, whether from Theranos or someone else. His thirty-five years of experience patenting medical inventions told him such a patent might eventually command up to $4 million for an exclusive license.

At 7:30 on the evening of Friday, September 23, 2005, Fuisz sent an email to his longtime patent attorney, Alan Schiavelli of the law firm Antonelli, Terry, Stout & Kraus, with the subject line "Blood Analysis—deviation from norm (individualized)":

> Al, Joe and I would like to patent the following application. It is a know [**sic**] art to check variou [**sic**] blood parameters like blood glucose, electrolytes, platelet activity, hematocrit etc. What we would like to cover as an improvement is the presence of a memory chip or other such storage device which could be programmed by a computer or similar device and contain the "normal parameters" for the individual patient. Thus if results would differ significantly from these norms—a notice would be given the user or health professional to repeat the sampling. If the significant difference persists on the retest, the device using existing technology well known in the art, to contact the physician, care center. [**sic**] pharma company or other or all.
>
> Please let me know next week if you could cover this. Thx. Rcf

Schiavelli was busy with other matters and didn't respond for several months. Fuisz finally got his attention on January 11, 2006, when he sent him another email saying he wanted to make a modification to his original idea: the alert mechanism would now be "a bar code or a radio tag label" on the package insert of the drug the patient was taking. A chip in the blood-testing device would scan the bar code and program the device to automatically send an alert to the patient's doctor if and when the patient's blood showed side effects from the drug.

Fuisz and Schiavelli exchanged more emails refining the concept, culminating in a fourteen-page patent application they filed with the U.S. Patent and Trademark Office on April 24, 2006. The proposed patent didn't purport to invent groundbreaking new technology. Rather, it combined existing ones—wireless data transmission, computer chips, and bar codes—into a physician alert mechanism that could be embedded in at-home blood-testing devices made by other companies. It made no secret of which particular company it was targeting: it mentioned Theranos by name in the fourth paragraph and quoted from its website.

Patent applications don't become public until eighteen months after they're filed, so neither Elizabeth nor her parents were initially aware of what Fuisz had done. Lorraine Fuisz and Noel Holmes continued to see each other regularly. The Holmeses settled into a new apartment they purchased on Wisconsin Avenue near the Naval Observatory. Lorraine drove over from McLean on several occasions and accompanied Noel, clad in her jogging suit, on walks through the neighborhood.

One day, Noel came over to the Fuisz home for lunch. Richard joined them out on the house's big stone patio and the conversation drifted to Elizabeth. She had just been profiled in **Inc.** magazine alongside several other young entrepreneurs, including Facebook's Mark Zuckerberg. The press her daughter was beginning to garner was a source of great pride to Noel.

As they nibbled on a meal Lorraine had picked up from a McLean gourmet shop, Fuisz suggested to Noel in a syrupy singsong voice he employed when he turned on the charm that he could be of assistance to Elizabeth. It was easy for a small company like Theranos to be taken advantage of by bigger ones, he noted. He didn't reveal his patent filing, but the comments may have been enough to put the Holmeses on alert. From that point on, interactions between the two couples became fraught.

The Fuiszes and Holmeses met twice for dinner in the waning months of 2006. One dinner was at Sushiko, a Japanese restaurant down the road from Chris and Noel's new apartment. Chris didn't eat much that evening. While visiting Elizabeth in Palo Alto, complications from a recent surgery had forced him to make a detour to Stanford Hospital. Fortunately, Elizabeth's boyfriend, Sunny, had arranged for him to stay in the hospital's VIP suite and covered the bill, he told the Fuiszes.

The conversation turned to Theranos, which had completed its second round of funding earlier in the year. Chris mentioned that the fund-raising had attracted some of the biggest investors in Silicon Valley, which was a good thing, he added, because he and Noel had put the $30,000 they'd saved for Elizabeth's Stanford tuition into the company.

The dinner then apparently grew testy for reasons that aren't entirely clear. Richard and Chris had never gotten along and Richard may have said something that got under the other man's skin. Whatever the case, according to Lorraine, Chris Holmes criticized the Chanel necklace she was wearing and later, after they'd settled the bill and wandered out onto Wisconsin Avenue, made what seemed like a veiled threat by bringing up the fact that John Fuisz, another one of Fuisz's sons from his first marriage, worked for his best friend. John Fuisz was indeed an attorney at the law firm McDermott Will & Emery, where Chris Holmes's closest friend, Chuck Work, was a senior partner.

Afterward, Noel and Lorraine's friendship began to fray. It had always been an odd pairing. Lorraine was originally from working-class Queens, a background betrayed by her coarse New York City accent. Noel, by contrast, was the epitome of the worldly Washington establishment woman. She'd spent part of her youth in Paris, when her father was assigned to the headquarters of the European Command.

In the following months, the two women got together for coffee several more times. But Chris Holmes, perhaps because he suspected that Richard Fuisz was up to something, always insisted on joining them, making their interactions awkward and tense. During one encounter, at Dean & DeLuca in Georgetown, the conversation became strained as they discussed the recent death of Lorraine's brother and the cat he'd left behind. Lorraine agonized about what to do with the cat, which seemed to exasperate Chris. He told her to just get rid of it and mimicked grabbing it and putting it in a bag. "The cat is not important," he said impatiently.

Since the Holmeses' move back to Washington, Noel had been going to the same hair salon as Lorraine in Tysons Corner, Virginia. They shared a hairdresser there named Claudia. As she was cutting Lorraine's hair one day, Claudia asked whether she and Noel were having problems. Noel had apparently been venting to Claudia. Embarrassed, Lorraine said she didn't want to talk about it and changed the subject.

Lorraine Fuisz and Noel Holmes saw each other one more time when Lorraine paid a visit to the Holmeses' apartment bearing cakes around Christmas 2007. Elizabeth, who was in town for the holidays, must have known that her parents and the Fuiszes were on the outs. She didn't say much and stole sidelong glances at her mother's friend.

Fuisz's patent application became available about

a week later, on January 3, 2008, to anyone who performed a search in the USPTO's online database. However, Theranos didn't learn of its existence for another five months until Gary Frenzel, the head of Theranos's chemistry team, came across it and called it to Elizabeth's attention. By then, the Holmeses and the Fuiszes were no longer on speaking terms and Fuisz was referring to his patent filing in conversations with his wife as "the Theranos killer."

THAT SUMMER, Chris Holmes went to see his old friend Chuck Work at the Washington offices of McDermott Will & Emery, two blocks east of the White House. Chris and Chuck were longtime friends. They'd met in 1971 when Chuck gave Chris a ride to an Army Reserve meeting. Although Chuck was five years older, they'd quickly realized they had a lot in common: they were both from California and had attended the same high school and college, the Webb Schools in Claremont, California, and Wesleyan University in Middletown, Connecticut.

Through the years, Chuck had often lent Chris a helping hand. After Enron collapsed, he let Chris use a visitor's office at his firm to conduct his job search. When Elizabeth's brother, Christian, had to leave St. John's high school in Houston because of what Chris described as a prank involving a film projector, Chuck was able to help Christian get into Webb because he served on the school's board. And when Elizabeth later dropped out of Stanford and needed help filing her first patent, Chuck put her in touch with colleagues at McDermott who specialized in that kind of work.

That was precisely the subject of Chris Holmes's visit on that summer day in 2008. Chris was agitated. He told Chuck someone named Richard Fuisz had stolen Elizabeth's idea and patented it. Fuisz, Chris noted pointedly, had a son who worked at McDermott named John. Chuck vaguely knew who John Fuisz was. Their paths had crossed once or twice at the firm when they'd overlapped on a case. He was also aware that McDermott had served as Theranos's patent attorneys for several years, since he was the one who'd made the initial introduction. But the rest of what Chris was saying was out of left field. He had no idea who Richard Fuisz was nor what patent he was referring to. As a favor to his old friend, he nevertheless agreed to see Elizabeth.

She came by a few weeks later, on September 22, 2008, and met with Chuck and another attorney named Ken Cage. Chuck had been McDermott's managing partner when the firm moved to the Robert A. M. Stern limestone building it occupied on Thirteenth Street, so he had the biggest and nicest corner office on the eighth floor. Elizabeth came in wheeling her blood-testing machine and sat down in one of two love seats placed catty-corner next to the office's big bay window. She didn't offer to demonstrate how the device worked, but Chuck thought it looked impressive at first glance. It was a big, shiny black-and-white cube with a digital touchscreen that bore a clear resemblance to an iPhone's.

Elizabeth got straight to the point. She wanted to know if McDermott would agree to represent Theranos against Richard Fuisz. Ken said they could look into filing a patent interference case if that's what she had in mind. Interference cases are contests adjudicated by the Patent and Trademark Office to determine who of two rival applicants vying to patent the same invention came up with it first. The winner's application gets priority even if it was filed later. Ken specialized in these types of cases.

Chuck was hesitant to do that, though. He told Elizabeth he would have to think about it and talk to some of his colleagues. Fuisz had a son who was a partner at the firm, which made the situation awkward, he said. Elizabeth didn't blink at the mention of John Fuisz. It was the opening she was waiting for. She asked whether it was possible that John had accessed confidential information from McDermott's Theranos file and leaked it to his father.

That seemed far-fetched to Chuck. It was the sort of thing that would get an attorney fired and disbarred. John was a patent litigator. He wasn't part of McDermott's separate patent prosecution team that drafted and filed patents. He had no reason or justification for accessing Theranos's file. Besides, he was a partner at the firm. Why would he commit career suicide? It didn't make sense. What's more, Theranos had moved all its patent work to the Silicon Valley law firm Wilson Sonsini two years earlier, in 2006. Chuck remembered Chris calling him and telling him apologetically at the time that Larry Ellison insisted Elizabeth use that firm. McDermott had obliged and transferred everything over to them. There was nothing for a McDermott attorney left to access.

After Elizabeth departed, Chuck consulted the heads of the firm's patent prosecution and patent litigation teams. The latter was John Fuisz's boss. He was told Theranos might have a plausible interference case against Richard Fuisz, but John Fuisz was a partner in good standing and the optics of the firm going up against the parent of one of its own partners were messy. Chuck decided to turn down Elizabeth's request. He informed her of his decision with a phone call a few weeks later. That was the last Chuck and McDermott expected to hear of the matter.

| SIX |

Sunny

Chelsea Burkett was burning out. It was the late summer of 2009 and she was working long hours at a Palo Alto startup, juggling what at a more established company would have been five different roles. Not that she was averse to hard work. Like most twenty-five-year-old Stanford graduates, striving was in her DNA. But she yearned for a little inspiration and she wasn't getting any from her job: Doostang, her employer, was a career website for finance professionals.

Chelsea had been one of Elizabeth's best friends at Stanford. As freshmen, they'd lived in adjacent dorms in Wilbur Hall, a big residential complex on the eastern edge of campus, and had immediately hit it off. Elizabeth wore a red-white-and-blue

"Don't mess with Texas" T-shirt and a big smile the first time they met. Chelsea found her sweet, smart, and fun.

Both were social and outgoing, with matching blue eyes. They did their share of drinking and partying and pledged a sorority, partly as a play for better housing. But, while Chelsea was a regular teenager still trying to find herself, Elizabeth seemed to know exactly who she wanted to be and what she wanted to do. When she returned to campus with a patent she'd written at the beginning of sophomore year, Chelsea was blown away.

The two young women had stayed in touch in the five years since Elizabeth had dropped out of school to launch Theranos. They didn't see each other often, but they texted occasionally. During one of these exchanges, Chelsea mentioned her job blues, prompting Elizabeth to write back, "Why don't you come work for me?"

Chelsea went to see Elizabeth at the Hillview Avenue office. It didn't take her friend long to sell her on Theranos. Elizabeth talked fervently about a future in which the company would save lives with its technology. It sounded a lot more interesting and noble to Chelsea than helping investment bankers find jobs. And Elizabeth was so persuasive. She had this intense way of looking at you while she spoke that made you believe in her and want to follow her.

They quickly settled on a role for Chelsea: she

would work in the client solutions group, which was in charge of setting up the validation studies Theranos was conducting to try to win pharmaceutical companies' business. Chelsea's first assignment would be to organize a study with Centocor, a division of Johnson & Johnson.

When she reported for her new job a few days later, Chelsea learned that she wasn't the only friend Elizabeth had hired. Just a week earlier, Ramesh "Sunny" Balwani had come on board as a senior Theranos executive. Chelsea had met Sunny once or twice but didn't know him well. She just knew he was Elizabeth's boyfriend and that they were living together in an apartment in Palo Alto. Elizabeth hadn't mentioned anything about Sunny joining the company, yet Chelsea now faced the reality of having to work with him. Or was it for him? She wasn't sure whether she reported to Sunny or to Elizabeth. Sunny's title, executive vice chairman, was both lofty and vague. Whatever his role was meant to be, he didn't waste any time asserting himself. From the get-go, he involved himself in every aspect of the company and became omnipresent.

Sunny was a force of nature, and not in a good way. Though only about five foot five and portly, he made up for his diminutive stature with an aggressive, in-your-face management style. His thick eyebrows and almond-shaped eyes, set above a mouth that drooped at the edges and a square chin, projected an air of menace. He was haughty and demeaning toward employees, barking orders and dressing people down.

Chelsea took an immediate dislike to him even though he made an effort to be nicer to her in deference to her friendship with Elizabeth. She didn't understand what her friend saw in this man, who was nearly two decades older than she was and lacking in the most basic grace and manners. All her instincts told her Sunny was bad news, but Elizabeth seemed to have the utmost confidence in him.

SUNNY HAD BEEN a presence in Elizabeth's life since the summer before she went to college. They'd met in Beijing in her third year attending Stanford's Mandarin program. Elizabeth had struggled to make friends that summer and gotten bullied by some of the students on the trip. Sunny, the lone adult among a group of college kids, had stepped in and come to her aid. That's how Elizabeth's mother, Noel, described the genesis of their relationship to Lorraine Fuisz.

Born and raised in Mumbai, Sunny first came to the United States in 1986 for his undergraduate studies. Afterward, he worked as a software engineer for a decade at Lotus and Microsoft. In 1999, he joined an Israeli entrepreneur named Liron Petrushka at a Santa Clara, California, startup called CommerceBid.com. Petrushka was developing a software program that would enable companies to pit their suppliers against one another in live online auctions to secure economies of scale and lower prices.

When Sunny joined CommerceBid, the dot-com frenzy was at its peak and the niche Petrushka's company was in, known as business-to-business e-commerce, had become red-hot. Analysts were breathlessly predicting that $6 trillion of commerce between corporations would soon be handled via the internet.

The sector's leader, Commerce One, had just gone public and seen its stock price triple on its first day of trading. It finished the year up more than 1,000 percent. That November, just a few months after Sunny was named CommerceBid's president and chief technology officer, Commerce One acquired the startup for $232 million in cash and stock. It was a breathtaking price for a company that had just three clients testing its software and barely any revenues. As the company's second-highest-ranking executive, Sunny pocketed more than $40 million. His timing was perfect. Five months later, the dot-com bubble popped and the stock market came crashing down. Commerce One eventually filed for bankruptcy.

Yet Sunny didn't see himself as lucky. In his mind, he was a gifted businessman and the Commerce One windfall was a validation of his talent. When Elizabeth met him a few years later, she had no reason to question that. She was an impressionable eighteen-year-old girl who saw in Sunny what she wanted to become: a successful and wealthy entrepreneur. He became her mentor, the person who would teach her about business in Silicon Valley.

It isn't clear exactly when Elizabeth and Sunny became romantically involved, but it appears to have been not long after she dropped out of Stanford. When they'd first met in China in the summer of 2002, Sunny was married to a Japanese artist named Keiko Fujimoto and living in San Francisco. By October 2004, he was listed as "a single man" on the deed to a condominium he purchased on Channing Avenue in Palo Alto. Other public records show Elizabeth moved into that apartment in July 2005.

Sunny spent the decade after his brief and lucrative stint at CommerceBid not doing much aside from enjoying his money and giving Elizabeth advice behind the scenes. He had stayed on at Commerce One as a vice president until January 2001 and then enrolled in business school at Berkeley. He later took classes in computer science at Stanford.

By the time he joined Theranos in September 2009, Sunny's legal record contained at least one red flag. To dodge taxes on his CommerceBid earnings, he'd hired the accounting firm BDO Seidman, which arranged for him to invest in a tax shelter. The maneuver generated an artificial tax loss of $41 million that offset his CommerceBid gains, all but eliminating his tax liability. When the Internal Revenue Service cracked down on the practice in 2004, Sunny was forced to pay the millions of dollars in back taxes he owed in a settlement with the agency. He turned around and sued BDO, claiming that he had been unsophisticated in tax matters and that the firm had knowingly misled him. The suit was settled on undisclosed terms in 2008.

Tax troubles aside, Sunny was proud of his wealth and liked to broadcast it with his cars. He drove a black Lamborghini Gallardo and a black Porsche 911. Both had vanity license plates. The one on the Porsche read "DAZKPTL" in mock reference to Karl Marx's treatise on capitalism. The Lamborghini's plate was "VDIVICI," a play on the phrase "Veni, vidi, vici" ("I came, I saw, I conquered"), which Julius Caesar used to describe his quick and decisive victory at the Battle of Zela in a letter to the Roman Senate.

The way Sunny dressed was also meant to telegraph affluence, though not necessarily taste. He wore white designer shirts with puffy sleeves, acid-washed jeans, and blue Gucci loafers. His shirts' top three buttons were always undone, causing his chest hair to spill out and revealing a thin gold chain around his neck. A pungent scent of cologne emanated from him at all times. Combined with the flashy cars, the overall impression was of someone heading out to a nightclub rather than to the office.

Sunny's expertise was software and that was where he was supposed to add value at Theranos. In one of the first company meetings he attended, he bragged that he'd written a million lines of code. Some employees thought that was preposterous. Sunny had worked at Microsoft, where teams of software engineers had written the Windows operating system at the rate of one thousand lines of code per year of development. Even if you assumed Sunny was twenty times faster than the Windows developers, it would still have taken him fifty years to do what he claimed.

Sunny was boastful and patronizing toward employees, but he was also strangely elusive at times. When Don Lucas showed up at the office once or twice a month to visit with Elizabeth, Sunny would suddenly vanish. One employee found a note on an office printer that Elizabeth had faxed to Lucas, in which she lauded Sunny's skills and résumé, so she hadn't concealed his hiring. But people like Dave Nelson, the engineer who had helped Tony Nugent build the first Edison prototype and who now sat across from Chelsea's cubicle, began to suspect that Elizabeth was downplaying to the board the breadth of Sunny's role.

There was also the murky question of what she told the board about their relationship. When Elizabeth informed Tony that Sunny was joining the company, Tony asked her point-blank whether they were still a couple. She responded that the relationship was over. Going forward, it was strictly business, she said. But that would prove not to be true.

CHELSEA'S CENTOCOR ASSIGNMENT took her to Antwerp, Belgium, in the fall of 2009. Daniel Young, a brainy bioengineering Ph.D. from MIT, accompanied her there. Daniel had been hired six months earlier to help add a new dimension to the Theranos blood-testing system: predictive modeling. When Elizabeth pitched pharmaceutical executives now, she told them that Theranos could forecast how patients would react to the drugs they were taking. Patients' test results would be input into a proprietary computer program the company had developed. As more results got fed into the program, its ability to predict how markers in the blood were likely to change during treatment would become better and better, she said.

It sounded cutting-edge, but there was a catch: the blood-test results had to be reliable for the computer program's predictions to have any value, and Chelsea started to have her doubts about that soon after she arrived in Belgium. Theranos was supposed to help Centocor assess how patients were responding to an asthma drug by measuring a biomarker in their blood called allergen-specific immunoglobulin E, or IgE, but the Theranos devices seemed very buggy to Chelsea. There were frequent mechanical failures. The cartridges either wouldn't slot into the readers properly or something inside the readers would malfunction. Even when the devices didn't break down, it could be a challenge coaxing any kind of output from them.

Sunny always blamed the wireless connection, and he was right in some instances. The process by which test results were generated involved a transatlantic round-trip of ones and zeros: when the blood test was completed, a cellular antenna on the reader beamed the voltage data produced by the light signal to a server in Palo Alto. The server analyzed the data and beamed back a final result to a cell phone in Belgium. When the cellular connection was weak, the data transmission would fail.

But there were other things besides the wireless connection that could interfere with the generation of a result. Nearly all blood tests require a certain amount of dilution to lower the concentration of substances in the blood that can wreak havoc on the test. In the case of chemiluminescent immunoassays—the class of tests the Edison performed—diluting the blood was necessary to filter out its light-absorbing pigments and other constituents that could interfere with the emission of the light signal. The amount of dilution the Theranos system required was greater than usual because of the small size of the blood samples Elizabeth insisted on. For the reader to have enough liquid to work with, the volume of the samples had to be increased significantly. The only way to do that was to dilute the blood more. And that in turn made the light signal weaker and harder to measure precisely. Put simply, some dilution was good, but too much dilution was bad.

The Edisons were also very sensitive to ambient temperature. To function properly, they needed to run at exactly 34 degrees Celsius. There were two 11-volt heaters built into the reader to try to maintain that temperature when a blood test was being run. But in colder settings, like certain hospitals in Europe, Dave Nelson had noticed that the little heaters didn't keep the readers warm enough.

Sunny didn't know or understand any of this because he had no background in medicine, much less laboratory science. Nor did he have the patience to listen to the scientists' explanations. It was easier to just blame the cellular connection. Chelsea wasn't much more knowledgeable about the science than Sunny was, but she was friendly with Gary Frenzel, the head of the chemistry team, and she gleaned from their conversations that the difficulties went far beyond connectivity issues.

What Chelsea didn't know at the time was that one of their pharmaceutical partners had already walked away from the startup. Earlier that year, Pfizer had informed Theranos that it was ending their collaboration because it was underwhelmed by the results of the Tennessee validation study. Elizabeth had tried to put the best spin she could on the fifteen-month study in a twenty-six-page report she'd sent to the New York pharmaceutical giant, but the report had betrayed too many glaring inconsistencies. The study had failed to show any clear link between drops in the patients' protein levels and the administration of the antitumor drugs. And the report had copped to some of the same snafus Chelsea was now witnessing in Belgium, such as mechanical failures and wireless transmission errors. It had blamed the latter on "dense foliage, metal roofs, and poor signal quality due to remote location."

Two of the Tennessee patients had called the Theranos offices in Palo Alto to complain that the readers wouldn't start because of temperature issues. "The solution," according to the report, had been to ask the patients to move the readers "away from A/C units and possible air currents." One patient had put the device in his RV and the other in a "very hot room" and the temperature extremes had "affected the readers' ability to maintain desired temperature," the report said.

The report was never shared with Chelsea. She didn't even know of the Pfizer study's existence.

WHEN SHE RETURNED to Palo Alto from her three-week stay in Antwerp, Chelsea discovered that Elizabeth and Sunny's attention had shifted from Europe to another part of the globe: Mexico.

A swine flu epidemic had been raging there since the spring and Elizabeth thought it offered a great opportunity to showcase the Edison.

The person who had planted that germ in her mind was Seth Michelson, Theranos's chief scientific officer. Seth was a math whiz who'd once worked in the flight simulator lab at NASA. His specialty was biomathematics, the use of mathematical models to help understand phenomena in biology. He was in charge of the predictive modeling efforts at Theranos and was Daniel Young's boss. Seth called to mind Doc Brown from the 1985 Michael J. Fox movie **Back to the Future**. He didn't have Doc's crazy white hair, but he sported a huge, frizzy gray beard that gave him a similar mad scientist look. Though in his late fifties, he still said "dude" a lot and became really animated when he was explaining scientific concepts.

Seth had told Elizabeth about a math model called SEIR (the letters stood for Susceptible, Exposed, Infected, and Resolved) that he thought could be adapted to predict where the swine flu virus would spread next. For it to work, Theranos would need to test recently infected patients and input their blood-test results into the model. That meant getting the Edison readers and cartridges to Mexico. Elizabeth envisioned putting them in the beds of pickup trucks and driving them to the Mexican villages on the front lines of the outbreak.

Chelsea was fluent in Spanish, so it was decided

that she would head down to Mexico with Sunny. Getting authorization to use an experimental medical device in a foreign country is usually no easy thing, but Elizabeth was able to leverage the family connections of a wealthy Mexican student at Stanford. He got Chelsea and Sunny an audience with high-ranking officials at the Mexican Social Security Institute, the agency that runs the country's public health-care system. IMSS approved the shipment of two dozen Edison readers to a hospital in Mexico City. The hospital, a sprawling facility called Hospital General de México, was located in Colonia Doctores, one of the city's most crime-ridden neighborhoods. Chelsea and Sunny were discouraged from going to and from the hospital on their own. A driver dropped them off inside the gates of the facility every morning and picked them up at the end of each day.

For weeks, Chelsea spent her days cooped up in a little room inside the hospital. The Edison readers were stacked on shelves along one wall. Refrigerators containing blood samples were lined up along another. The blood came from infected patients who'd been treated at the hospital. Chelsea's job was to warm up the samples, put them in the cartridges, slot the cartridges into the readers, and see if they tested positive for the virus.

Once again, things did not go smoothly. Frequently, the readers flashed error messages, or the result that came back from Palo Alto was negative for the virus when it should have been positive. Some of the readers didn't work at all. And Sunny continued to blame the wireless transmission.

Chelsea grew frustrated and miserable. She questioned what she was even doing there. Gary Frenzel and some of the other Theranos scientists had told her that the best way to diagnose H1N1, as the swine flu virus was called, was with a nasal swab and that testing for it in blood was of questionable utility. She'd raised this point with Elizabeth before leaving, but Elizabeth had brushed it off. "Don't listen to them," she'd said of the scientists. "They're always complaining."

Chelsea and Sunny had several meetings with IMSS officials at the Mexican health ministry to update them on their work. Sunny didn't speak or understand a word of Spanish, so Chelsea did all the talking. As the meetings dragged on, Sunny's face would betray a mixture of annoyance and concern. Chelsea suspected he was worried she was telling the Mexicans that the Theranos system didn't work. She enjoyed seeing him squirm.

Back in Palo Alto, word around the office was that Elizabeth was negotiating a deal to sell four hundred Edison readers to the Mexican government. The deal was supposed to bring in a much-needed influx of cash. The $15 million Theranos had raised in its first two funding rounds was long gone and the company had already burned through the $32 million Henry Mosley had been instrumental in bringing in during its Series C round in late 2006. The company was being kept afloat with a loan Sunny had personally guaranteed.

Meanwhile, Sunny was also traveling to Thailand to set up another swine flu testing outpost. The epidemic had spread to Asia, and the country was one of the region's hardest hit with tens of thousands of cases and more than two hundred deaths. But unlike in Mexico, it wasn't clear that Theranos's activities in Thailand were sanctioned by local authorities. Rumors were circulating among employees that Sunny's connections there were shady and that he was paying bribes to obtain blood samples from infected patients. When a colleague of Chelsea's in the client solutions group named Stefan Hristu quit immediately upon returning from a trip to Thailand with Sunny in January 2010, many took it to mean the rumors were true.

Chelsea was back from Mexico by then and the Thailand gossip spooked her. She knew there was an anti-bribery law called the Foreign Corrupt Practices Act. Violating it was a felony that could result in prison time.

WHEN SHE STOPPED to think about it, there were a lot of things that made Chelsea uncomfortable about Theranos. And none more so than Sunny. He spawned a culture of fear with his intimidating behavior. Firings had always been a common occurrence at the company, but in late 2009 and early 2010 it was Sunny who took on the role of hatchet man. Chelsea even learned a new expression: to disappear someone. That's how employees used the normally intransitive verb when someone was dismissed. "Sunny disappeared him," they would say, conjuring up the image of a Mafia hit in 1970s Brooklyn.

The scientists, especially, were afraid of Sunny. One of the only ones who stood up to him was Seth Michelson. A few days before Christmas, Seth had gone out and purchased polo shirts for his group. Their color matched the green of the company logo and they had the words "Theranos Biomath" emblazoned on them. Seth thought it was a nice team-building gesture and paid for it out of his own pocket.

When Sunny saw the polos, he got angry. He didn't like that he hadn't been consulted and he argued that Seth's gift to his team made the other managers look bad. Earlier in his career, Seth had worked at Roche, the big Swiss drugmaker, where he'd been in charge of seventy people and an annual budget of $25 million. He decided he wasn't going to let Sunny lecture him about management. He pushed back and they got into a yelling match.

After that, Sunny seemed to have it in for Seth and frequently harassed him, which led Seth to look for another job. He found one a few months later at a company based in Redwood City called Genomic

Health and walked into Elizabeth's office, resignation letter in hand, to give his notice. Sunny, who was there, opened up the letter, read it, then threw it back in Seth's face.

"I won't accept this!" he shouted.

Seth shouted back, deadpan, "I have news for you, sir: in 1863, President Lincoln freed the slaves."

Sunny's response was to throw him out of the building. It was weeks before Seth was able to retrieve his math books, scientific journals, and the pictures of his wife on his desk. He had to enlist the company's new lawyer, Jodi Sutton, and a security guard to help him pack his things late on a weeknight when Sunny wasn't around.

Sunny also got into it with Tony Nugent one Friday evening. He'd been giving direct orders and putting intense pressure on a young engineer on Tony's team, causing him to fall apart from the stress. Tony confronted Sunny about it and their argument quickly escalated. Working himself into a fury, Sunny yelled that he was doing everyone a favor by volunteering his time to the company and people should be a little more appreciative.

"I've made enough money to look after my family for seven generations. I don't need to be here!" he screamed in Tony's face.

Tony roared back in his Irish brogue, "I don't have a cent and I don't need to be here either!"

Elizabeth had to step in to defuse the situation. Dave Nelson thought that Tony would be fired and that he'd have a new boss by Monday morning. Yet Tony somehow survived the confrontation.

Chelsea tried to complain to Elizabeth about Sunny, but she couldn't get through to her. Their bond seemed too strong to be shaken. Whenever Elizabeth came out of her office, which was separated from Sunny's by a glass conference room, he would immediately pop out of his and walk with her. Often, he accompanied her all the way to the bathrooms in the back of the building, prompting some employees to wonder half jokingly if they were snorting lines of cocaine back there.

By February 2010, after six months on the job, Chelsea had lost all her enthusiasm for working at Theranos and was thinking of quitting. She hated Sunny. The Mexico and Thailand projects seemed to be losing steam as the swine flu pandemic subsided. The company was lurching from one ill-conceived initiative to another like a child with attention deficit disorder. On top of it all, Chelsea's boyfriend lived in Los Angeles and she was flying back and forth between L.A. and the Bay Area every weekend to see him. The commute was killing her.

As she debated what to do, something happened that hastened her decision. One day, the Stanford student whose family connections Elizabeth had tapped in Mexico came by with his father. Chelsea wasn't there to witness the visit, but the office was buzzing about it afterward. The father was going through some sort of cancer scare. Upon hearing of his health worries, Elizabeth and Sunny had convinced him to let Theranos test his blood for cancer biomarkers. Tony Nugent, who wasn't there for the encounter either, heard about it later that day from Gary Frenzel.

"Well, that was interesting," Gary told Tony, his voice conveying bewilderment. "We played doctor today."

Chelsea was appalled. The validation study in Belgium and the experiments in Mexico and Thailand were one thing. Those were supposed to be for research purposes only and to have no bearing on the way patients were treated. But encouraging someone to rely on a Theranos blood test to make an important medical decision was something else altogether. Chelsea found it reckless and irresponsible.

She became further alarmed when not long afterward Sunny and Elizabeth began circulating copies of the requisition forms doctors used to order blood tests from laboratories and speaking excitedly about the great opportunities that lay in consumer testing.

I'm done, Chelsea thought to herself. This has crossed too many lines.

She approached Elizabeth and told her she wanted to resign but decided to keep her qualms to herself. Instead, she told her friend that her weekend commutes were taking too great a toll and that she wanted to move to Los Angeles full-time, which in any case was true. She offered to stay on for a transition period, but Elizabeth and Sunny didn't want her to. If Chelsea was leaving, better she do so right away, they told her. They asked her not to say anything to the three employees who reported to her on her way out. Chelsea protested. It didn't feel right to flee like a thief in the middle of the night. But Sunny and Elizabeth were firm: she was not to speak to them.

Chelsea walked out of the building and into the Palo Alto sunshine with conflicting emotions. The dominant one was relief. But she also felt bad that she hadn't been able to say goodbye to her team and to tell them why she was leaving. She would have given them the official reason—that she was moving to L.A.—but Sunny and Elizabeth hadn't trusted her to do that. They'd wanted to control the narrative of her departure.

Chelsea also worried about Elizabeth. In her relentless drive to be a successful startup founder, she had built a bubble around herself that was cutting her off from reality. And the only person she was letting inside was a terrible influence. How could her friend not see that?

| SEVEN |

Dr. J

As the calendar turned from 2009 to 2010, America remained mired in a deep economic malaise. Over the previous two years, nearly 9 million people had lost their jobs in the worst downturn since the Great Depression. Millions more had been hit with foreclosure notices. But in the 1,500-square-mile area south of San Francisco that forms the boundaries of Silicon Valley, animal spirits were stirring again.

A new luxury hotel on Sand Hill Road called the Rosewood was always full, despite room rates that reached a thousand dollars a night. With its imported palm trees and proximity to the Stanford campus, it had quickly become the destination of choice for venture capitalists, startup founders, and out-of-town investors who flocked to its restaurant and poolside bar to discuss deals and be seen. Bentleys, Maseratis, and McLarens lined its stone parking lot.

While the rest of the country licked its wounds from the devastating financial crisis, a new technology boom was getting under way, fueled by several factors. One of them was the wild success of Facebook. In June 2010, the social network's private valuation rose to $23 billion. Six months later, it jumped to $50 billion. Every startup founder in the Valley wanted to be the next Mark Zuckerberg and every VC wanted a seat on the next rocket ship to riches. The emergence of Twitter, which was valued at more than $1 billion in late 2009, added to the excitement.

Meanwhile, the iPhone and competing smartphones featuring Google's Android operating system were beginning to usher in a shift to mobile computing, as cellular networks became faster and capable of handling larger amounts of data. Wildly popular mobile games like **Angry Birds,** which millions of iPhone users were paying a dollar each to download, seeded the notion that you could build a business around a smartphone app. In the spring of 2010, an obscure startup called UberCab did a beta launch of its black car hailing service in San Francisco.

All of this might not have been enough to ignite the new boom, however, if it hadn't been for another key ingredient: rock-bottom interest rates. To rescue the economy, the Federal Reserve had slashed rates to close to zero, making traditional investments like bonds unattractive and sending investors searching for higher returns elsewhere. One of the places they turned to was Silicon Valley.

Suddenly, the managers of East Coast hedge funds that normally invested only in publicly traded stocks were making the pilgrimage West in search of promising new opportunities in the private startup world. They were joined by executives from old, established companies looking to harness the Valley's innovation to rejuvenate businesses battered by the recession. Among this latter group was a sixty-five-year-old man from Philadelphia who greeted people with high fives in lieu of handshakes and went by the sobriquet "Dr. J."

Dr. J's real name was Jay Rosan and he was in fact a doctor, though he had spent most of his career working for big corporations. He was a member of Walgreens's innovation team, which was tasked with identifying new ideas and technologies that could reboot growth at the 109-year-old drugstore chain. Dr. J operated out of an office in the Philadelphia suburb of Conshohocken that Walgreens had inherited from its 2007 acquisition of Take Care Health Systems, an operator of in-store clinics where he'd previously been employed.

In January 2010, Theranos had approached Walgreens with an email stating that it had developed small devices capable of running any blood test from a few drops pricked from a finger in real time and for less than half the cost of traditional laboratories. Two months later, Elizabeth and Sunny traveled to Walgreens's headquarters in the Chicago suburb of Deerfield, Illinois, and gave a presentation to a group of Walgreens executives. Dr. J, who flew up from Pennsylvania for the meeting, instantly recognized the potential of the Theranos technology. Bringing the startup's machines inside Walgreens stores could open up a big new revenue stream for the retailer and be the game changer it had been looking for, he believed.

It wasn't just the business proposition that appealed to Dr. J. A health nut who carefully watched his diet, rarely drank alcohol, and was fanatical about getting a swim in every day, he was passionate about empowering people to live healthier lives. The picture Elizabeth presented at the meeting of making blood tests less painful and more widely available so they could become an early warning system against disease deeply resonated with him. That evening, he could barely contain his excitement over dinner at a wine bar with two Walgreens colleagues who weren't privy to the secret discussions with Theranos. After asking them to keep what he was about to tell them confidential, he revealed in a hushed tone that he'd found a company he was convinced would change the face of the pharmacy industry.

"Imagine detecting breast cancer before the mammogram," he told his enraptured colleagues, pausing for effect.

A FEW MINUTES BEFORE eight a.m. on August 24, 2010, a group of rental cars pulled up in front of 3200 Hillview Avenue in Palo Alto. A stocky man with glasses and dimples on his wide nose stepped out of one of them. His name was Kevin Hunter and he headed a small lab consulting firm called Colaborate. He was part of a Walgreens delegation led by Dr. J that had flown to California for a two-day meeting with Theranos. The drugstore chain had hired him a few weeks before to help evaluate and set up a partnership it was negotiating with the startup.

Hunter had a special affinity for the business Walgreens was in: his father, grandfather, and great-grandfather had all been pharmacists. Growing up, he'd spent the summers helping his dad man the counter and stock the shelves of the pharmacies he ran on air bases in New York, Texas, and New Mexico. As familiar as he was with drugstores, though, Hunter's real expertise was with clinical laboratories. After getting his MBA at the University of Florida, he had spent the first eight years of his career working for Quest Diagnostics, the giant provider of lab services. He had subsequently launched Colaborate, which advised clients ranging from hospitals to private equity firms about laboratory issues.

The first thing Hunter noticed as he shut the door of his rental car and walked toward the entrance of Theranos's office was a shiny black Lamborghini parked right next to it. Looks like someone is trying to impress us, he thought.

Elizabeth and Sunny greeted him and the rest of the Walgreens team at the top of a flight of stairs and showed them to the glass conference room between their offices. They were joined there by Daniel Young, who had succeeded Seth Michelson as head of Theranos's biomath team. On the Walgreens side, in addition to Hunter and Dr. J, three others had made the trip: a Belgian executive named Renaat Van den Hooff, a financial executive named Dan Doyle, and Jim Sundberg, who worked with Hunter at Colaborate.

Dr. J high-fived Sunny and Elizabeth, then sat down and kicked off the meeting with the same line he always used when he introduced himself: "Hi, I'm Dr. J and I used to play basketball." Hunter had already heard him use it a dozen times in the few weeks they'd worked together and no longer thought it was funny, but for Dr. J it was a joke that never seemed to grow old. It elicited a few awkward chuckles.

"I'm so excited that we're doing this!" Dr. J then exclaimed. He was referring to a pilot project the companies had agreed to. It would involve placing

Theranos's readers in thirty to ninety Walgreens stores no later than the middle of 2011. The stores' customers would be able to get their blood tested with just a prick of the finger and receive their results in under an hour. A preliminary contract had already been signed, under which Walgreens had committed to prepurchase up to $50 million worth of Theranos cartridges and to loan the startup an additional $25 million. If all went well with the pilot, the companies would aim to expand their partnership nationwide.

It was unusual for Walgreens to move this quickly. Opportunities the innovation team identified usually got waylaid in internal committees and slowed down by the retailer's giant bureaucracy. Dr. J had managed to fast-track this one by going straight to Wade Miquelon, Walgreens's chief financial officer, and getting him behind the project. Miquelon was due to fly in that evening and join them at the next day's session.

About half an hour into discussions centering on the pilot, Hunter asked where the bathroom was. Elizabeth and Sunny visibly stiffened. Security was paramount, they said, and anyone who left the conference room would have to be escorted. Sunny accompanied Hunter to the bathroom, waited for him outside the bathroom door, and then walked him back to the conference room. It seemed to Hunter unnecessary and strangely paranoid.

On his way back from the bathroom, he scanned the office for a laboratory but didn't see anything that looked like one. That's because it was downstairs, he was told. Hunter said he hoped to see it at some point during the visit, to which Elizabeth responded, "Yes, if we have time."

Theranos had told Walgreens it had a commercially ready laboratory and had provided it with a list of 192 different blood tests it said its proprietary devices could handle. In reality, although there was a lab downstairs, it was just an R&D lab where Gary Frenzel and his team of biochemists conducted their research. Moreover, half of the tests on the list couldn't be performed as chemiluminescent immunoassays, the testing technique the Edison system relied on. They required different testing methods beyond the Edison's scope.

The meeting resumed and stretched into the middle of the afternoon, at which point Elizabeth suggested they grab an early dinner in town. As they got up from their chairs, Hunter asked again to see the lab. Elizabeth tapped Dr. J on the shoulder and motioned for him to follow her outside the conference room. He returned moments later and told Hunter it wasn't going to happen. Elizabeth wasn't willing to show them the lab yet, he said. Instead, Sunny showed the Walgreens team his office. There was a sleeping bag on the floor behind his desk, his bathroom had a shower in it, and he kept a change of clothes on hand. He worked such long hours that on many nights he crashed at the office, he proudly told the visitors.

As they headed out to eat, Sunny and Elizabeth made them leave at staggered intervals. They didn't want everyone to arrive at the restaurant at the same time on the grounds that it risked attracting notice. They also instructed Hunter and his colleagues not to use names. When Hunter got to the restaurant, a little sushi place on El Camino Real called Fuki Sushi, the hostess took him to a private room in the back with sliding doors where Elizabeth was waiting.

The cloak-and-dagger theatrics struck Hunter as silly. It was four in the afternoon and the restaurant was empty. There was no one to conceal their presence from. What's more, if there was anything likely to draw attention, it was Sunny's Lamborghini in the parking lot.

Hunter was beginning to grow suspicious. With her black turtleneck, her deep voice, and the green kale shakes she sipped on all day, Elizabeth was going to great lengths to emulate Steve Jobs, but she didn't seem to have a solid understanding of what distinguished different types of blood tests. Theranos had also failed to deliver on his two basic requests: to let him see its lab and to demonstrate a live vitamin D test on its device. Hunter's plan had been to have Theranos test his and Dr. J's blood, then get retested at Stanford Hospital that evening and compare the results. He'd even arranged for a pathologist to be on standby at the hospital to write the order and draw their blood. But Elizabeth claimed she'd been given too little notice even though he'd made the request two weeks ago.

There was something else that bothered Hunter: Sunny's attitude. He acted both superior and cavalier. When the Walgreens side had broached bringing its IT department in on the pilot preparations, Sunny had dismissed the idea out of hand by saying, "IT are like lawyers, avoid them as long as possible." That kind of approach sounded to Hunter like a recipe for problems.

Dr. J didn't seem to share his skepticism, though. He appeared taken with Elizabeth's aura and to revel in the Silicon Valley scene. He reminded Hunter of a groupie who'd flown across the country to attend a concert played by his favorite band.

When they reconvened at the Theranos office the next morning, they were joined by Wade Miquelon, the Walgreens CFO. Wade had negotiated the pilot contract directly with Elizabeth. He too seemed to be a big fan of hers. Midway through that day's meeting, Elizabeth made a big show of giving Miquelon an American flag that she said had been flown over a battlefield in Afghanistan. She'd written a dedication to Walgreens on it.

Hunter thought the whole thing was bizarre. Walgreens had brought him here to vet Theranos's technology, but he hadn't been allowed to do so.

The only thing they had to show for their visit was an autographed flag. And yet, Dr. J and Miquelon didn't seem to mind. As far as they were concerned, the visit had gone swimmingly.

A MONTH LATER, in September 2010, a group of Walgreens executives met with Elizabeth and Sunny in a conference room at the drugstore chain's Deerfield headquarters. The mood was festive. Red balloons with the Walgreens logo floated above a table laden with hors d'oeuvres. Wade Miquelon and Dr. J were unveiling "Project Beta," the code name for the Theranos pilot, to senior Walgreens executives.

Standing in front of a slide titled "Disrupting the Lab Industry" projected on a big screen, one of the Walgreens executives was singing along to "Imagine." To celebrate the alliance, the innovation team had come up with the idea of adapting the lyrics to the John Lennon song and using it as the partnership's anthem. When the awkward karaoke act was over, Elizabeth and Sunny encouraged the Walgreens executives to get their blood tested. They had brought along several black-and-white machines to the meeting. The Walgreens executives lined up to get their fingers pricked behind Kermit Crawford, the president of the pharmacy business, and Colin Watts, the head of the innovation team.

Hunter, who was now working for Walgreens

full-time as an onsite consultant for the innovation team, didn't take part in the meeting. But when he heard that several Walgreens executives had had their blood tested, he figured this was an opportunity to finally see how the technology performed. He told himself to follow up with Elizabeth about the test results next time they talked. In a report he'd put together after the Palo Alto visit, Hunter had warned that Theranos might be "overselling or overstating . . . where they are at scientifically with the cartridges/devices." He'd also recommended that Walgreens embed someone at Theranos through the pilot's launch and had volunteered one of his Colaborate colleagues, a petite British woman by the name of June Smart who'd recently completed a stint administering Stanford's labs, for the assignment. Theranos had rejected the idea.

Hunter asked about the blood-test results a few days later on the weekly video conference call the companies were using as their primary mode of communication. Elizabeth responded that Theranos could only release the results to a doctor. Dr. J, who was dialed in from Conshohocken, reminded everyone that he was a trained physician, so why didn't Theranos go ahead and send him the results? They agreed that Sunny would follow up separately with him.

A month passed and still no results.

Hunter's patience was wearing thin. During that

week's call, the two sides discussed a sudden change Theranos had made to its regulatory strategy. It had initially represented that its blood tests would qualify as "waived" under the Clinical Laboratory Improvement Amendments, the 1988 federal law that governed laboratories. CLIA-waived tests usually involved simple laboratory procedures that the Food and Drug Administration had cleared for home use.

Now, Theranos was changing its tune and saying the tests it would be offering in Walgreens stores were "laboratory-developed tests." It was a big difference: laboratory-developed tests lay in a gray zone between the FDA and another federal health regulator, the Centers for Medicare and Medicaid Services. CMS, as the latter agency was known, exercised oversight of clinical laboratories under CLIA, while the FDA regulated the diagnostic equipment that laboratories bought and used for their testing. But no one closely regulated tests that labs fashioned with their own methods. Elizabeth and Sunny had a testy exchange with Hunter over the significance of the change. They maintained that all the big laboratory companies mostly used laboratory-developed tests, which Hunter knew not to be true.

To Hunter, the switch made it all the more important to check the accuracy of Theranos's tests. He suggested doing a fifty-patient study in which they would compare Theranos results to ones from

Stanford Hospital. He'd done work with Stanford and knew people there; it would be easy to arrange. On the computer screen, Hunter noticed an immediate change in Elizabeth's body language. She became visibly guarded and defensive.

"No, I don't think we want to do that at this time," she said, quickly changing the subject to other items on the call's agenda.

After they hung up, Hunter took aside Renaat Van den Hooff, who was in charge of the pilot on the Walgreens side, and told him something just wasn't right. The red flags were piling up. First, Elizabeth had denied him access to their lab. Then she'd rejected his proposal to embed someone with them in Palo Alto. And now she was refusing to do a simple comparison study. To top it all off, Theranos had drawn the blood of the president of Walgreens's pharmacy business, one of the company's most senior executives, and failed to give him a test result!

Van den Hooff listened with a pained look on his face.

"We can't not pursue this," he said. "We can't risk a scenario where CVS has a deal with them in six months and it ends up being real."

Walgreens's rivalry with CVS, which was based in Rhode Island and one-third bigger in terms of revenues, colored virtually everything the drugstore chain did. It was a myopic view of the world that was hard to understand for an outsider like Hunter who wasn't a Walgreens company man. Theranos had cleverly played on this insecurity. As a result, Walgreens suffered from a severe case of FoMO—the fear of missing out.

Hunter pleaded with Van den Hooff to at least let him peek inside the one black-and-white reader Theranos had left with them after the Project Beta kickoff party. He was dying to tear the strips of security tape off its case and crack it open. Theranos had sent them some test kits for it, but they were for obscure blood tests like a "flu susceptibility" panel that no other lab he knew of offered. It was therefore impossible to compare their results to anything. How convenient, Hunter had noted. Moreover, the kits were expired.

Van den Hooff said no. In addition to signing confidentiality agreements, they'd been sternly warned not to tamper with the reader. The contract the companies had signed stated that Walgreens agreed "not to disassemble or otherwise reverse engineer the Devices or any component thereof."

Trying to contain his frustration, Hunter made one last request. Theranos always invoked two things as proof that its technology had been vetted. The first was the clinical trial work it did for pharmaceutical companies. Documents it gave Walgreens stated that the Theranos system had been "comprehensively validated over the last seven years by ten of the largest fifteen pharma companies." The second was a review of its technology Dr. J had supposedly commissioned from Johns Hopkins University's medical school.

Hunter had placed calls to pharmaceutical companies and hadn't been able to get anyone on the phone to confirm what Theranos was claiming, though that was hardly proof of anything. He now asked Van den Hooff to show him the Johns Hopkins review. After some hesitation, Van den Hooff reluctantly handed him a two-page document.

When Hunter was done reading it, he almost laughed. It was a letter dated April 27, 2010, summarizing a meeting Elizabeth and Sunny had had with Dr. J and five university representatives on the Hopkins campus in Baltimore. It stated that they had shown the Hopkins team "proprietary data on test performance" and that Hopkins had deemed the technology "novel and sound." But it also made clear that the university had conducted no independent verification of its own. In fact, the letter included a disclaimer at the bottom of the second page: "The materials provided in no way signify an endorsement by Johns Hopkins Medicine to any product or service."

Hunter told Van den Hooff the letter was meaningless. Judging from the Belgian's expression, he sensed that he was starting to make headway. Van den Hooff's confidence seemed shaken. Hunter knew that Dan Doyle, the executive responsible for the innovation team's finances, shared some of his skepticism. If he could convert Van den Hooff to their point of view, they might get Dr. J and Wade Miquelon to see the light and avert a potential disaster.

WALGREENS WASN'T the only big retail partner Theranos was courting. During this same period, Theranos employees began noticing visits to the Hillview Avenue office from an older gentleman with an earnest air who wore rimless glasses and a suit and tie. It was Steve Burd, the CEO of Safeway.

Burd had been at the helm of Safeway, one of the country's biggest supermarket chains, for seventeen years. Along the way, his disciplined focus on the X's and O's of the grocery business, which had earned him plaudits from Wall Street during his first decade as CEO, had given way to an intense interest in health care.

He'd gotten hooked on the subject after realizing that Safeway's rising medical costs threatened to someday bankrupt the company if he didn't do something to tame them. He'd pioneered innovative wellness and preventive health programs for his employees and become an advocate for universal health coverage, making him one of the only Republican CEOs to embrace many of the tenets of Obamacare. Like Dr. J, he was serious about his own health. He worked out on a treadmill at five every morning and lifted weights in the evenings after dinner.

At Burd's invitation, Elizabeth came to the supermarket chain's headquarters in Pleasanton, on the other side of San Francisco Bay, to make a presentation. As the Safeway CEO and a group of his top executives listened intrigued, she described how her phobia of needles had led her to develop breakthrough technology that made blood tests not only more convenient, but faster and cheaper. She brought along one of her black-and-white devices to demonstrate how it worked.

The presentation had a strong impact on Larree Renda, Safeway's executive vice president. Renda's husband was battling lung cancer. His blood had to be tested frequently so that doctors could adjust his drug regimen. Each blood draw was an exercise in torture because his veins were collapsing. Theranos's fingerprick system would be a godsend for him, she thought.

Renda, who had started out at Safeway at age sixteen as a part-time bagger and worked her way up the corporate ladder to become one of Burd's most trusted executives, could see that her boss was also very impressed. The Theranos proposition dovetailed perfectly with his wellness philosophy and offered a way to improve the supermarket chain's stagnating revenues and razor-thin profit margins.

Before long, Safeway too signed a deal with Theranos. Under the agreement, it loaned the startup $30 million and pledged to undertake a massive renovation of its stores to make room for sleek new clinics where customers would have their blood tested on the Theranos devices.

Burd was over the moon about the partnership. He saw Elizabeth as a precocious genius and treated her with rare deference. Normally loath to leave his office unless it was absolutely necessary, he made an exception for her, regularly driving across the bay to Palo Alto. On one occasion, he arrived bearing a huge white orchid. On another, he brought her a model of a private jet. Her next one, he predicted, would be real. Burd was aware of Theranos's parallel discussions with Walgreens. Elizabeth told him his company would be the exclusive purveyor of Theranos blood tests in supermarkets, while Walgreens would be granted exclusivity in drugstores. Neither company was thrilled with the arrangement, but both saw it as better than missing out on a huge new business opportunity.

BACK IN CHICAGO, Hunter's efforts to get Van den Hooff to take his suspicions seriously were dashed in mid-December 2010 when Van den Hooff informed his colleagues that he would be leaving at the end of the year. He'd been offered a job as the CEO of a company in New Jersey that made temperature indicators for pharmaceutical companies. It was a career opportunity he couldn't pass up.

Walgreens appointed an internal replacement, a female executive named Trish Lipinski who had some exposure to the laboratory world. Before coming to Walgreens, she had worked at the College of American Pathologists, the medical association representing laboratory scientists. Hunter wasted no time letting her know how he felt about the Theranos project. "I've got to stop this because someday this is going to be a black eye on someone," he told her.

He also voiced his skepticism directly to Dr. J, but that was of little use. Dr. J was a staunch and tireless advocate for Theranos. If anything, he thought Walgreens was moving too slowly. After learning of the model jet Steve Burd had given Elizabeth, he'd railed to Trish that Walgreens needed to show her more love. To Hunter's amazement, he had even stopped asking Elizabeth and Sunny about the test results from the kickoff party. He was apparently willing to let Theranos get away with not producing them.

Dr. J had a powerful ally in Wade Miquelon. A sharp dresser with a taste for expensive suits and designer eyeglasses, Wade was gregarious and well liked at Walgreens. However, many of his colleagues had begun to question his judgment after a story in the **Chicago Tribune** revealed that he'd been arrested for driving drunk that fall for the second time in a little over a year. He shouldn't have been behind the wheel of a car at all: his driver's license was still suspended from the previous arrest. To make matters worse, he'd refused to take a

Breathalyzer and failed field sobriety tests. The incident earned him a new nickname in the hallways of Walgreens headquarters: Michelob.

Wade's DUIs and Dr. J's blind cheerleading for Theranos didn't inspire confidence that Project Beta was in the best hands. But that was beyond Hunter's purview. He focused on what he could control, continuing to ask tough questions on the weekly video calls until one day in early 2011 Lipinski told him that Elizabeth and Sunny no longer wanted him on the calls or in meetings between the companies. They felt that he was creating too much tension and that it interfered with getting work done, she said. Walgreens had no choice but to comply or Theranos would walk away, she added.

Hunter tried to convince her to rebuff the demand. Why was Walgreens paying his firm $25,000 a month to look out for its best interests if it was going to keep him at arm's length and make it harder for him to do his job? It made no sense. His protestations were politely ignored and Elizabeth and Sunny got their way. Hunter continued to work with the innovation team and to provide his expertise when asked, but his exclusion from subsequent calls and meetings marginalized him and limited his input.

In the meantime, Walgreens pushed ahead with the project. As part of the pilot preparations, Hunter joined the innovation team on a field trip to an unmarked warehouse in an industrial park a few miles from the Deerfield campus. Inside, the company had built a full-scale replica of one of its stores. It featured a blood-testing laboratory, with shelves designed specifically to accommodate the dimensions of the black-and-white Theranos readers.

Seeing the mock store and its little lab brought home to Hunter how real it all was. Soon, actual patients were going to get their blood drawn and tested in one of these, he thought uneasily.

| EIGHT |

The miniLab

With Walgreens and Safeway on board as retail partners, Elizabeth suddenly faced a problem of her own making: she had told both companies her technology could perform hundreds of tests on small blood samples. The truth was that the Edison system could only do immunoassays, a type of test that uses antibodies to measure substances in the blood. Immunoassays included some commonly ordered lab tests such as tests to measure vitamin D or to detect prostate cancer. But many other routine blood tests, ranging from cholesterol to blood sugar, required completely different laboratory techniques.

Elizabeth needed a new device, one that could perform more than just one class of test. In November 2010, she hired a young engineer named Kent Frankovich and put him in charge of designing it. Kent had just obtained a master's degree in mechanical engineering from Stanford. Before that, he'd spent two years working for NASA's Jet Propulsion Laboratory in Pasadena where he'd helped build Curiosity, the Mars rover. Kent in turn recruited Greg Baney, a friend he'd met at NASA who'd gone on to work for SpaceX, Elon Musk's Los Angeles–based rocket company. At six feet five and 260 pounds, Greg was built like an NFL lineman, but his physique belied a sharp intellect and a keen sense of observation.

For a period of several months, Kent and Greg became Elizabeth's favorite employees. She sat in on their brainstorming sessions and made suggestions about what robotics systems they should consider using. She gave them company credit cards and let them charge whatever equipment and supplies they wanted.

Elizabeth christened the machine she assigned them to build the "miniLab." As its name suggested, her overarching concern was its size: she still nurtured the vision of someday putting it in people's homes and wanted something that could fit on a desk or a shelf. This posed engineering challenges because, in order to run all the tests she wanted, the miniLab would need to have many more components than the Edison. In addition to the Edison's photomultiplier tube, the new device would need to cram three other laboratory instruments in one small space: a spectrophotometer, a cytometer, and an isothermal amplifier.

None of these were new inventions. The first commercial spectrophotometer was developed in 1941 by the American chemist Arnold Beckman, founder of the lab equipment maker Beckman Coulter. It works by beaming rays of colored light through a blood sample and measuring how much of the light the sample absorbs. The concentration of a molecule in the blood is then inferred from the level of light absorption. Spectrophotometers are used to measure substances like cholesterol, glucose, and hemoglobin. Cytometry, a way of counting blood cells, was invented in the nineteenth century. It's used to diagnose anemia and blood cancers, among other disorders.

Laboratories all over the world had been using these instruments for decades. In other words, Theranos wasn't pioneering any new ways to test blood. Rather, the miniLab's value would lie in the miniaturization of existing lab technology. While that might not amount to groundbreaking science, it made sense in the context of Elizabeth's vision of taking blood testing out of central laboratories and bringing it to drugstores, supermarkets, and, eventually, people's homes.

To be sure, there were already portable blood analyzers on the market. One of them, a device that looked like a small ATM called the Piccolo Xpress, could perform thirty-one different blood tests and produce results in as little as twelve minutes. It required only three or four drops of blood for a panel of a half dozen commonly ordered tests. However, neither the Piccolo nor other existing portable analyzers could do the entire range of laboratory tests. In Elizabeth's mind, that was going to be the miniLab's selling point.

Greg spent a lot of time studying commercial instruments made by diagnostics equipment makers to reverse engineer them and make them smaller. He ordered a spectrophotometer from a company called Ocean Optics and broke it apart to understand how it worked. It turned into an interesting project, but it made him question their approach.

Instead of building new instruments from scratch to fit the arbitrary dimensions Elizabeth had laid out, Greg felt they would do better to take the off-the-shelf components they were laboring to miniaturize and integrate them together to test how the overall system worked. Once they had a working prototype, they could then worry about shrinking it. Emphasizing the system's size first and how it worked later was putting the cart before the horse. But Elizabeth wouldn't budge.

Greg was in the midst of a breakup with the girl he'd dated in L.A., so he came to the office on Saturdays to get his mind off it. He could see that Elizabeth really appreciated that. She saw it as a sign of loyalty and dedication. She told Greg she wanted to see Kent come in on weekends too; it bothered her that his friend didn't. Keeping a work-life balance seemed a foreign concept to her. She was at work all the time.

Like most people, Greg had been taken aback by Elizabeth's deep voice when he'd first met her. He soon began to suspect it was affected. One evening, as they wrapped up a meeting in her office shortly after he joined the company, she lapsed into a more natural-sounding young woman's voice. "I'm really glad you're here," she told him as she got up from her chair, her pitch several octaves higher than usual. In her excitement, she seemed to have momentarily forgotten to turn on the baritone. When Greg thought about it, there was a certain logic to her act: Silicon Valley was overwhelmingly a man's world. The VCs were all male and he couldn't think of any prominent female startup founder. At some point, she must have decided the deep voice was necessary to get people's attention and be taken seriously.

A few weeks after the voice incident, Greg picked up another clue that Theranos wasn't your usual workplace. He had become friendly with Gary Frenzel. Although Gary looked like a slob—he weighed three hundred pounds and walked around the office in baggy jeans, an oversized T-shirt, and Crocs—Greg found him to be one of the smartest people at the company. Gary had a bad case of sleep apnea and, more than once during meetings,

Greg had watched him doze off only to suddenly snap awake to refute a dumb idea someone had put forward and suggest a brilliant alternative.

As they walked out of the office together one day, Gary lowered his voice and in a conspiratorial tone told Greg something that startled the younger man: Elizabeth and Sunny were in a romantic relationship. Greg felt blindsided. He thought it was inappropriate for the CEO of a company and its number-two executive to be sleeping together, but what bothered him more was the fact they were hiding it. This was a crucial piece of information that he felt should have been disclosed to new recruits. For Greg, the revelation cast everything about Theranos in a new light: If Elizabeth wasn't being forthright about that, what else might she be lying about?

NEPOTISM AT THERANOS took on a new dimension in the spring of 2011 when Elizabeth hired her younger brother, Christian, as associate director of product management. Christian Holmes was two years out of college and had no clear qualifications to work at a blood diagnostics company, but that mattered little to Elizabeth. What mattered far more was that her brother was someone she could trust.

Christian was a handsome young man with eyes the same deep shade of blue as his sister's, but that was where the similarities between them began and ended. Christian had none of his sister's ambition and drive; he was a regular guy who liked to watch sports, chase girls, and party with friends. After graduating from Duke University in 2009, he'd worked as an analyst at a Washington, D.C., firm that advised corporations about best practices.

When he first arrived at the company, Christian didn't have much to do, so he spent part of his days reading about sports. He hid it by cutting and pasting articles from the ESPN website into empty emails so that, from afar, it looked like he was absorbed in work-related correspondence. Christian soon recruited four of his fraternity brothers from Duke: Jeff Blickman, Nick Menchel, Dan Edlin, and Sani Hadziahmetovic. They were later joined by a fifth Duke friend, Max Fosque. They rented a house together near the Palo Alto country club and became known inside Theranos as "the Frat Pack." Like Christian, none of the other Duke boys had any experience or training relevant to blood testing or medical devices, but their friendship with Elizabeth's brother vaulted them above most other employees in the company hierarchy.

By then, Greg had convinced several of his own friends to join Theranos. Two of them were buddies from his undergraduate days at Georgia Tech, Jordan Carr and Ted Pasco. The third was a friend he'd made in Pasadena while working for NASA named Trey Howard. Trey happened to have gone to college at Duke a few years before the Frat Pack.

Jordan, Trey, and Ted were all assigned to the product management group with Christian and his friends, but they weren't granted the same level of access to sensitive information. Many of the hush-hush meetings Elizabeth and Sunny held to strategize about the Walgreens and Safeway partnerships were off limits to them, whereas Christian and his fraternity brothers were invited in.

The Frat Pack endeared itself to Sunny and Elizabeth by working long days. Sunny was constantly questioning employees' commitment to the company—the number of hours a person put in at the office, whether he or she was doing productive work or not, was his ultimate gauge of that commitment. At times, he would sit in the big glass conference room and stare out at the rows of cubicles trying to identify who was slacking off.

The numerous late nights they spent at the office left no time for exercise, so Christian and his friends snuck workouts in during the day. To elude Sunny's watchful gaze, they ducked out of the building at different times using different exits. They were also careful never to return at the same time or together. Ted Pasco, who had left a career on Wall Street to try his luck in Silicon Valley but didn't have any clear duties during his first few months at Theranos, amused himself by timing their exits and entries.

Several members of the Frat Pack joined Greg and two of his colleagues from the engineering department for lunch on the big terrace overlooking the parking lot one day. A discussion about the low IQs of some of the world's top soccer players led them to debate the question, Would you rather be smart and poor or dumb and rich? The three engineers all chose smart and poor, while the Frat Pack voted unanimously for dumb and rich. Greg was struck by how clearly the line was drawn between the two groups. They were all in their mid- to late twenties with good educations, but they valued different things.

Christian and his friends were always ready and willing to do Elizabeth and Sunny's bidding. Their eagerness to please was on display when news broke that Steve Jobs had died on the evening of October 5, 2011. Elizabeth and Sunny wanted to pay Jobs a tribute by flying an Apple flag at half-mast on the grounds of the Hillview Avenue building. The next morning, Jeff Blickman, a tall redhead who'd played varsity baseball at Duke, volunteered for the mission. He couldn't locate any suitable Apple flag for sale, so Blickman had one custom made out of vinyl. It featured the famous Apple logo in white against a black background. The store he went to took a while to make it. Blickman didn't return with it until late in the day. In the meantime, work at the company came to a standstill as Elizabeth and Sunny moped around the office, consumed by the hunt for the Apple flag.

Greg had been aware of Elizabeth's fascination with Jobs. She referred to him as "Steve" as if they were close friends. At one point, she'd told him that a documentary espousing a 9/11 conspiracy theory wouldn't have been available on iTunes if "Steve" hadn't believed there was something to it. Greg thought that was silly. He was pretty sure Jobs hadn't personally screened all the movies for rent or sale on iTunes. Elizabeth seemed to have this exaggerated image of him as an all-seeing and all-knowing being.

A month or two after Jobs's death, some of Greg's colleagues in the engineering department began to notice that Elizabeth was borrowing behaviors and management techniques described in Walter Isaacson's biography of the late Apple founder. They were all reading the book too and could pinpoint which chapter she was on based on which period of Jobs's career she was impersonating. Elizabeth even gave the miniLab a Jobs-inspired code name: the 4S. It was a reference to the iPhone 4S, which Apple had coincidentally unveiled the day before Jobs passed away.

GREG'S HONEYMOON PERIOD at Theranos ended when his sister applied for a job at the company.

After interviewing with both Elizabeth and Sunny in April 2011, she received an offer to join the product management team the following month but decided to turn it down and stay with her employer, the accounting firm PwC. The next day, a Saturday, Greg was at the office working. Elizabeth was there too but wouldn't acknowledge his presence, which he found odd since she usually made a point to, especially on weekends. The following week, Greg stopped being invited to her brainstorming sessions with Kent. It dawned on him that she'd taken his sister's decision personally and that he was now paying the price for it.

Not long after, a chill descended on Kent's own relationship with Elizabeth. For all intents and purposes, Kent was the chief architect of the miniLab. A talented engineer who loved to build stuff, he was also dabbling with a side project in his spare time: bicycle lights that lit up both wheels and the road, providing improved visibility and safety for the rider at night. He'd pitched the concept on Kickstarter and, much to his surprise, was able to raise $215,000 in forty-five days. It was the seventh-largest sum raised on the crowdfunding platform that year. What had been a hobby suddenly looked like it could become a viable business.

Kent told Elizabeth about his successful Kickstarter campaign, thinking she wouldn't mind. But he badly miscalculated: she and Sunny were furious. They viewed it as a major conflict of interest and asked him to transfer his bike-lights patent to Theranos. The paperwork Kent had signed when he joined the company entitled them to any intellectual property he produced while employed there, they contended. Kent disagreed. He'd worked on his little venture during his free time and felt he had done nothing wrong. He also failed to see how a new type of bicycle light posed a threat to a maker of blood-testing equipment. But Elizabeth and Sunny wouldn't let it go. In meeting after meeting, they tried to get him to turn over the patent. They ratcheted up the pressure by bringing Theranos's new senior counsel, David Doyle, to some of the meetings.

As he watched the standoff unfold, Greg became convinced that it wasn't so much about the patent as it was about punishing Kent for his perceived disloyalty. Elizabeth expected her employees to give their all to Theranos, especially ones like Kent whom she entrusted with big responsibilities. Not only had Kent not given his all, he'd devoted part of his time and energy to another engineering project. It explained why he hadn't been coming in on weekends like she wanted him to. As she saw it, Kent had betrayed her. In the end, a fragile compromise was reached: Kent would go on a leave of absence to give his bicycle-light venture a shot. When he was done indulging his pet project, they'd have a conversation about whether, and under what conditions, he could return.

Kent's departure put Elizabeth in a foul mood. She now looked to Greg and others to pick up the slack. Greg also sensed a growing urgency in Elizabeth and Sunny's behavior. They seemed to be squeezing the engineering team to meet some sort of deadline without communicating to them what that deadline was. They must have promised someone something, he thought.

As Elizabeth grew impatient with the pace of the miniLab's development, Greg bore the brunt of her frustration. When the engineering team gathered for weekly status updates, she opened the meetings by staring at him silently without blinking until he broke the ice with a polite "Hello Elizabeth, how are you today?" He began keeping detailed notes of what was discussed and agreed to at each meeting that he could refer back to the following week to keep emotions out of it.

Several times, Elizabeth came downstairs to the engineers' workshop and hovered over Greg while he worked. He politely acknowledged her, then resumed working in silence. It was some sort of strange power play and he was determined not to get rattled by it.

One afternoon, Elizabeth called him into her office and told him she sensed cynicism emanating from him. After a long silence in which he debated telling her she was right, Greg decided to keep his growing disenchantment to himself and told a fib: he was upset because Sunny had rejected several job applicants that he thought were well qualified and hoped the company would hire.

Elizabeth must have believed him because she relaxed noticeably. "You need to tell us about these things," she said.

ON A WEEKDAY EVENING in December 2011, Theranos chartered several buses to transport its employees, which now numbered more than one hundred, to the Thomas Fogarty Winery in Woodside. It was Elizabeth's favorite place to hold corporate events. The winery's main building and its adjacent events facility were built on stilts into the hillside and offered panoramic views of the estate's rolling vineyards and of the Valley beyond.

The occasion was the company's annual Christmas party. As employees sipped drinks from an open bar inside the winery's main building before sitting down to dinner, Elizabeth gave a speech.

"The miniLab is the most important thing humanity has ever built. If you don't believe this is the case, you should leave now," she declared, scanning her audience with a dead serious look on her face. "Everyone needs to work as hard as humanly possible to deliver it."

Trey, the friend Greg had met while living in Pasadena and recruited to Theranos, tapped Greg's foot. They glanced at each other knowingly. What Elizabeth had just said confirmed their armchair psychoanalysis of their boss: she saw herself as a world historical figure. A modern-day Marie Curie.

Six weeks later, they were back at the Fogarty Winery, this time to celebrate the Safeway alliance. Standing on the deck of the open-air events house, Elizabeth harangued employees for forty-five minutes as the fog rolled in, like General Patton addressing his troops before the Allied landings. The sweeping view before them was appropriate, she said, because Theranos was about to become Silicon Valley's dominant company. Toward the end she boasted, "I'm not afraid of anything," adding after a brief pause, "except needles."

By this point, Greg had become fully disillusioned and resolved to stick around only two more months until his stock options vested on the first anniversary of his hiring. He'd recently gone to a job fair at his alma mater, Georgia Tech, and had found himself unable to talk the company up to students who stopped by the Theranos booth. Instead, he'd focused his advice on the merits of a career in Silicon Valley.

Part of the problem was that Elizabeth and Sunny seemed unable, or unwilling, to distinguish between a prototype and a finished product. The miniLab Greg was helping build was a prototype, nothing more. It needed to be tested thoroughly and fine-tuned, which would require time. A lot of time. Most companies went through three cycles of prototyping before they went to market with a product. But Sunny was already placing orders for components to build one hundred miniLabs, based on a first, untested prototype. It was as if Boeing built one plane and, without doing a single flight test, told airline passengers, "Hop aboard."

One of the difficulties that would need to be resolved through extensive testing was thermal. When you packed that many instruments into a small, enclosed space, you introduced unanticipated variations in temperature that could interfere with the chemistry and throw off the performance of the overall system. Sunny seemed to think that if you just put all the parts in a box and turned it on, it would work. If only it were that easy.

At one point, he pulled Greg and an older engineer named Tom Brumett into the big glass conference room and questioned their passion. Greg prided himself on never losing his cool, but this time he did. He leaned menacingly over the conference table. His huge, muscular frame towered over Sunny.

"God damn it, we are working our asses off," he growled.

Sunny backed off and apologized.

SUNNY WAS a tyrant. He fired people so often that it gave rise to a little routine in the warehouse downstairs. John Fanzio, the affable supply-chain manager, worked down there, and it had become the trusted place where employees came to vent or gossip. Every few days, Edgar Paz, the head of Theranos's security team, would come down with a mischievous look on his face, a badge hidden in his hand. At the sight of him, John and the logistics team would gather in excitement, knowing what was coming. As Paz drew closer, he would slowly spin the badge from its necklace and reveal the face on the front, eliciting gasps of surprise. It was Sunny's latest victim.

John had become good friends with Greg, Jordan, Trey, and Ted. Together, the five of them formed a little island of sanity at the company. John was probably the only strategic supply-chain manager in the Bay Area who worked just feet away from the cold roll-up door of the loading dock, but he liked it because it kept him away from Sunny's scrutiny and his obsessive focus on the number of hours people worked.

Unfortunately, working in the warehouse is what eventually brought about John's own demise. One morning in February 2012, one of the receiving guys who worked there with him arrived at work in a shiny new Acura. He proudly showed it to John, who complimented him on it. The next day, though, the car had a big dent in it. Someone had hit it in the office parking lot. John found the culprit by checking all the other cars in the lot for signs of a collision. It belonged to one of the Indian consultants Sunny had brought in to help with software development.

John confronted the owner when he came outside for a smoke break with his friends. He denied it even though John had used a tape measure to match the size of the dent on the Acura to the scrape on his car, a trick he'd learned from watching cops do it. John advised his warehouse colleague to report the accident to the police and show them the evidence. That's when the situation escalated. The Indian software consultants went upstairs to complain to Sunny, who came down in such a fury that his hands were visibly shaking.

"Oh really, you want to be a cop?" Sunny yelled at John, his voice dripping with sarcasm. "Go be a cop!"

He then turned to one of the security guards who was standing nearby and, motioning toward John, told him, "Get him out of here." After watching Edgar Paz playfully reveal the identity of scores of employees Sunny had fired over the previous year, it was John's turn to get the boot.

His friend's firing didn't sit well with Greg, and solidified his resolve to leave the company. A month later, a young engineer he worked with inadvertently fried some miniLab electrical boards. Sunny summoned Greg and Tom Brumett to his office and angrily demanded that they tell him who was to blame. They refused, knowing full well that Sunny would fire the young man if they gave him his name.

As it happened, Greg's stock options had just vested. Later in the day, he returned to Sunny's office and handed him his letter of resignation. Sunny calmly accepted it but, as soon as Greg left, he summoned Trey, Jordan, and Ted one after the other to gauge their intentions. All three assured him that Greg's decision didn't affect them and that they remained committed to working at Theranos for the long haul, knowing that was what Sunny wanted to hear.

Greg worked one last Saturday during his notice period. Sunny was grateful and invited him to a meeting Elizabeth was holding the following Monday in Newark, a small city directly across San Francisco Bay from Palo Alto. Theranos had just leased a huge manufacturing facility there to produce the miniLab in large quantities. Elizabeth was unveiling the cavernous, empty space to employees. She caught sight of Greg in the audience as she spoke and locked her gaze on him.

"If anyone here believes you are not working on the best thing humans have ever built or if you're cynical, then you should leave," she said, reprising the themes of her Christmas speech. Then, while continuing to look directly at Greg, she singled out Trey, Jordan, and Ted for special praise. There were some 150 employees assembled and she could have called out the names of any one of them, but she chose to commend the three people she knew were his friends. It was a final public rebuke.

IN THE MONTHS after Greg left, the revolving door at Theranos continued to swing at a furious pace. One of the more surreal incidents involved a burly software engineer named Del Barnwell. Big Del, as people called him, was a former Marine helicopter pilot. Sunny was on his case about not working long-enough hours. He'd gone as far as to review security footage to track Big Del's comings and goings and confronted him in a meeting in his office, claiming the tapes showed he worked only eight hours a day. "I'm going to fix you," Sunny told him, as if Del were a broken toy.

But Big Del didn't want to be fixed. Shortly after the meeting, he emailed his resignation notice to Elizabeth's assistant. He heard nothing back and dutifully worked the last two weeks of his notice period. Then, at four p.m. on a Friday, Big Del picked up his belongings and walked toward the building's exit. Sunny and Elizabeth suddenly came running down the stairs behind him. He couldn't leave without signing a nondisclosure agreement, they said.

Big Del refused. He'd already signed a confidentiality agreement when he was hired and, besides, they'd had two weeks to schedule an exit interview with him. Now he was free to go as he pleased and he damn well intended to. As he pulled out of the parking lot in his yellow Toyota FJ Cruiser, Sunny sent a security guard after him to try to stop him. Big Del ignored the guard and drove off.

Sunny called the cops. Twenty minutes later, a police cruiser quietly pulled up to the building with its lights off. A highly agitated Sunny told the officer that an employee had quit and departed with company property. When the officer asked what he'd taken, Sunny blurted out in his accented English, "He stole property in his mind."

| NINE |

The Wellness Play

Safeway's business was doing poorly. The supermarket chain had just announced a 6 percent drop in its profits for the last three months of 2011, a disappointing performance its longtime CEO Steve Burd was struggling to explain to the dozen analysts who had dialed in to the company's quarterly earnings call.

One of them, Ed Kelly from the Swiss bank Credit Suisse, was gently needling Burd for using stock buybacks to mask the bad results. By reducing the number of shares it had outstanding, buybacks could artificially raise a company's earnings per share—the headline number investors focused on—even if its actual earnings fell. It was an old trick that astute Wall Street analysts versed in corporate sleights of hand saw right through.

Piqued, Burd said he disagreed. He was confident that Safeway's fortunes were about to improve, which would make purchasing its own shares a smart investment. To justify his optimism, he cited three initiatives the company was pursuing. The first two were shrugged off as old news by his hard-to-please audience, but the analysts' ears pricked up when he got to the third.

"We're contemplating a significant . . . huh—I'm just going to call it a wellness play," he said cryptically.

It was the first time Burd had made any public mention of this. He didn't elaborate, but the message the analysts took away was that the stodgy, ninety-seven-year-old grocery chain had a secret plan to jump-start its stagnating business. Inside Safeway, this secret plan was code-named "Project T-Rex." It referred to none other than the company's partnership with Theranos, which by now—February 2012—was two years in the making.

Burd had high hopes for the venture. He'd ordered the remodeling of more than half of Safeway's seventeen hundred stores to make room for upscale clinics with deluxe carpeting, custom wood cabinetry, granite countertops, and flat-screen TVs. Per Theranos's instructions, they were to be called wellness centers and had to look "better than a spa." Although Safeway was shouldering the entire cost of the $350 million renovation on its own, Burd expected it to more than pay for itself once the new clinics started offering the startup's novel blood tests.

A few weeks after the earnings call, Burd and his executive team took a group of analysts on a tour of a Safeway store a few miles from his home in the scenic San Ramon Valley east of Oakland. The analysts were shown the store's new wellness center, but Burd remained evasive about what sort of service it was going to offer. Even the store's manager was in the dark. Theranos had insisted on absolute secrecy until the launch.

There had been quite a few delays since the companies had first agreed to do business together. At one point, Elizabeth told Burd that the earthquake that struck eastern Japan in March 2011 was interfering with Theranos's ability to produce the cartridges for its devices. Some Safeway executives found the excuse far-fetched, but Burd accepted it at face value. He was starry-eyed about the young Stanford dropout and her revolutionary technology, which fit so perfectly with his passion for preventive health care.

Elizabeth had a direct line to Burd and answered only to him. A war room had been set up in the Pleasanton headquarters where a small group of Safeway executives privy to Project T-Rex met once a week to discuss its progress. Burd attended all the meetings, either in person or via conference call if he was traveling. When questions or issues came up that had to be taken back to Theranos, he would pipe up with what became a refrain: "I'll talk to Elizabeth about it." Larree Renda, the executive who had started out at Safeway as a teenage bagger in 1974 and climbed the corporate ranks to become one of Burd's top deputies, and other executives involved in the project were surprised by how much latitude he gave the young woman. He usually held his deputies and the company's business partners to firm deadlines, but he allowed Elizabeth to miss one after the other. Some of Burd's colleagues knew he had two sons. They began to wonder if he saw in Elizabeth the daughter he'd never had. Whatever it was, he was in her thrall.

AFTER ALL THE DELAYS, the partnership seemed to finally be getting off the ground in the early months of 2012: as a beta run before a full launch, the companies had agreed that Theranos would take over the blood testing at an employee health clinic Safeway had opened on its corporate campus in Pleasanton. The clinic was part of Burd's strategy to curb the supermarket operator's health-care costs by encouraging its workers to take better care of themselves. It offered free checkups. Employees who scored well on them were entitled to discounts on their health plan premiums. Conveniently located next to the gym on the Safeway campus, it was staffed with a doctor and three nurse practitioners and featured five exam rooms. It also had a little lab. A new sign in the reception area read, "Testing done by Theranos."

The employee clinic was part of Renda's portfolio. Among other responsibilities, she oversaw Safeway Health, the subsidiary Burd had created to sell the retailer's health benefits expertise to other companies. Renda's husband had lost his battle with lung cancer since Elizabeth had first showed up in Pleasanton two years before, but she hoped Theranos's painless finger-stick tests would spare others the torment he'd endured getting repeatedly stuck with needles in the last months of his life.

Renda had just hired Safeway's first chief medical officer. His name was Kent Bradley and he came from the U.S. Army, where he had served for more than seventeen years after attending West Point and the armed forces' medical school in Bethesda, Maryland. Bradley's last military assignment had been to run the European wing of Tricare, the health insurance program for active and retired servicemen. Among other responsibilities, Renda gave the soft-spoken former army doctor oversight of the campus clinic.

Bradley had worked with a lot of sophisticated medical technologies in the army, so he was curious to see the Theranos system in action. However, he was surprised to learn that Theranos wasn't planning on putting any of its devices in the Pleasanton clinic. Instead, it had stationed two phlebotomists there to draw blood, and the samples they collected were couriered across San Francisco Bay to Palo Alto for testing. He also noticed that the phlebotomists were drawing blood from every employee twice, once with a lancet applied to the index finger and a second time the old-fashioned way with a hypodermic needle inserted in the arm. Why the need for venipunctures—the medical term for needle draws—if the Theranos finger-stick technology was fully developed and ready to be rolled out to consumers, he wondered.

Bradley's suspicions were further aroused by the amount of time it took to get results back. His understanding had been that the tests were supposed to be quasi-instantaneous, but some Safeway employees were having to wait as long as two weeks to receive their results. And not every test was performed by Theranos itself. Even though the startup had never said anything about outsourcing some of the testing, Bradley discovered that it was farming out some tests to a big reference laboratory in Salt Lake City called ARUP.

What really set off Bradley's alarm bells, though, was when some otherwise healthy employees started coming to him with concerns about abnormal test results. As a precaution, he sent them to get retested at a Quest or LabCorp location. Each time, the new set of tests came back normal, suggesting the Theranos results were off. Then one day, a senior

Safeway executive got his PSA result back. The acronym stands for "prostate-specific antigen," which is a protein produced by cells in the prostate gland. The higher the protein's concentration in a man's blood, the likelier he is to have prostate cancer. The senior Safeway executive's result was very elevated, indicating he almost certainly had prostate cancer. But Bradley was skeptical. As he had done with the other employees, he sent his worried colleague to get retested at another lab and, lo and behold, that result came back normal too.

Bradley put together a detailed analysis of the discrepancies. Some of the differences between the Theranos values and the values from the other labs were disturbingly large. When the Theranos values did match those of the other labs, they tended to be for tests performed by ARUP.

Bradley shared his concerns with Renda and with Brad Wolfsen, the president of Safeway Health. Her faith already shaken by the delays of the past two years, Renda encouraged him to talk to Burd about them, which Bradley did. But Burd politely brushed him off, assuring the ex–army doctor that the Theranos technology had been vetted and was sound.

THE BLOOD SAMPLES DRAWN from Safeway employees in Pleasanton were being couriered to a one-story building with a stone façade on East Meadow

Circle in Palo Alto. Theranos had temporarily set up its fledgling lab there in the spring of 2012 while it moved the rest of its growing operations from Hillview Avenue to a larger building nearby formerly occupied by Facebook.

A few months earlier, the lab had obtained a certificate attesting that it was in compliance with CLIA, the federal law that governed clinical laboratories, but such certificates weren't difficult to obtain. Although the ultimate enforcer of CLIA was the Centers for Medicare and Medicaid Services, the federal agency delegated most routine lab inspections to states. In California, they were handled by the state department of health's Laboratory Field Services division, which an audit had shown to be badly underfunded and struggling to fulfill its oversight responsibilities.

Had Steve Burd been allowed inside the East Meadow Circle lab, a network of rooms located in the center of the low-slung building, he would have noticed that it didn't contain a single Theranos proprietary device. That's because the miniLab was still under development and nowhere near ready for patient testing. What the lab did contain was more than a dozen commercial blood and body-fluid analyzers made by companies such as Chicago-based Abbott Laboratories, Germany's Siemens, and Italy's DiaSorin. The lab was run by an awkward pathologist named Arnold Gelb, who went by Arne (pronounced "Arnie"), and staffed with a handful of clinical laboratory scientists, or CLSs—lab technicians who are certified by the state to handle human samples. Although it made use of only commercial instruments at this juncture, there were still plenty of things that could and did go wrong.

The main problem was the lab's dearth of experienced personnel. One of the CLSs, a fellow by the name of Kosal Lim, was so sloppy and poorly trained that one of his counterparts, Diana Dupuy, was convinced he was jeopardizing the accuracy of test results. Dupuy was from Houston and had trained at MD Anderson, the city's world-renowned cancer center. She'd spent most of her seven years since becoming a CLS as a blood-transfusion specialist, which had given her extensive exposure to CLIA regulations. She operated strictly by the book and didn't hesitate to report violations when she saw them.

To Dupuy, Lim's blunders were inexcusable. They included ignoring manufacturers' instructions for how to handle reagents; putting expired reagents in the same refrigerator as current ones; running patient tests on lab equipment that hadn't been calibrated; improperly performing quality-control runs on an analyzer; doing tasks he hadn't been trained to do; and contaminating a bottle of Wright's stain, a mixture of dyes used to differentiate blood cell types. Dupuy, who had a fiery streak, confronted Lim on several occasions, telling him at one point that she was going to become a laboratory inspector to root out bad lab technicians like him. When he proved unable to meet her standards, she began documenting his poor practices in regular emails she sent to Gelb and to Sunny, often attaching photos to prove her point.

Dupuy also had concerns about the competence of the two phlebotomists Theranos had stationed in Pleasanton. Blood is typically spun down in a centrifuge before it's tested to separate its plasma from the patient's blood cells. The phlebotomists hadn't been trained to use the centrifuge they'd been given and didn't know how long or at what speed to spin down patients' blood. When they arrived in Palo Alto, the plasma samples were often polluted with particulate matter. She also discovered that many of the blood-drawing tubes Theranos was using were expired, making the anticoagulant in them ineffective and compromising the integrity of the specimens.

Shortly after making one of her complaints, Dupuy was sent to Delaware to train on a new Siemens analyzer Theranos had purchased. When she returned from her trip a week later, she noticed that the lab was spotless. Sunny, who appeared to have been waiting for her, summoned her into a meeting room. In an intimidating tone, he informed her that he had taken a tour of the lab in her absence and found not a single one of her complaints to be justified. He then brought up the fact that she had allowed her boyfriend into the building to help her carry out her luggage on the day she'd flown to Delaware. It was a serious breach of the company's security policy and he had decided to fire her for it, he said. After letting that sink in for a moment, he called Gelb in and asked him whether he valued Dupuy as a member of the lab and wanted to keep her on. Gelb said he did, at which point Sunny grudgingly reversed himself. Dupuy was still employed after all.

Shaken, she went back to her desk in a daze. The next thing she knew, an employee from the IT department tapped her on the shoulder and asked her to step out into the hallway with him. He was trying to reestablish her company cell-phone connection and needed some information from her. Before changing his mind, Sunny had ordered it disconnected along with her company email and her access to the corporate network.

Someone as outspoken as Dupuy was bound not to last long at Theranos. Three weeks later, on a Friday morning, Sunny returned to the East Meadow Circle building and fired her again, this time for good. She was immediately escorted out of the building without being given the chance to gather her personal belongings. The reason for her dismissal was that she had called attention to the fact that one of the lab's main vendors had put its purchase orders on hold because of unpaid bills.

Upset about the way she was treated, Dupuy fired

off an email to Sunny that weekend insisting that she be allowed to retrieve her belongings, which in addition to lab books included a makeup bag containing her eyeglasses and her California CLS license. The email, on which she copied Elizabeth, provided a searing indictment of Sunny's management style and of the state of the lab:

> I was warned by more than 5 people that you are a loose cannon and it all depends on your mood as to how [**sic**] what will trigger you to explode. I was also told that anytime someone deals with you it's never a good outcome for that person.
>
> . . .
>
> The CLIA lab is in trouble with Kosal running the show and no one watching him or Arne. You have a mediocre Lab Director taking up for a sub-par CLS for whatever reasons. I fully guarantee that Kosal will certainly make a huge mistake one day in the lab that will adversely affect patient results. I actually think he has already done this on several accounts but has put the blame on the reagents. Just as you stated everything he touches is a disaster!
>
> I am [**sic**] only hope that somehow I bring awareness to you that you have created a work environment where people hide things from you out of fear. You cannot run a company through fear and intimidation . . . it will only work for a period of time before it collapses.

Sunny agreed to have someone meet her in front of the East Meadow Circle building to return her belongings but warned her that she would be hearing from the company's lawyers. Over the next several days, Dupuy received a series of sternly worded emails from David Doyle, the Theranos senior counsel, demanding that she sign a declaration pledging to return to Theranos or "permanently destroy" any materials from her employment at the company and abide by her confidentiality obligations.

Dupuy initially refused and hired an Oakland attorney to threaten the company with a wrongful termination lawsuit, but her lawyer advised her to back down and to sign the document once Theranos brought in a high-powered attorney from Wilson Sonsini. Going up against Silicon Valley's premier law firm was a losing battle, he told her. She reluctantly followed his advice.

SAFEWAY OF COURSE had no knowledge of any of this. It continued to let Theranos handle the blood testing at its Pleasanton clinic throughout 2012 and into 2013. It also began hiring phlebotomists to staff the wellness centers it had built in dozens of its stores in Northern California. But as the months went by, Theranos continued to push back the date for a launch.

Burd was asked about the status of Safeway's

mysterious "wellness play" on its first-quarter earnings call in late April 2012. He replied that it wasn't yet "ready for prime time" but that when the company did unveil it, it would "have a material impact" on its financial results. In the next earnings call in July, he volunteered that it would play "out in all likelihood in the fourth quarter." However, the fourth quarter came and went without a launch.

By this point, some Safeway executives were getting angry. They were being denied their bonuses because the company was missing its financial targets, which had factored in the anticipated extra revenues and profits from the Theranos partnership. Matt O'Rell, an executive in Safeway's finance department, had been tasked with coming up with revenue projections for the wellness centers. Working from the aggressive assumption that each of them would attract an average of fifty patients per day, he had forecast $250 million in extra revenue per year. Not only had that revenue failed to materialize, Safeway had spent $100 million more than that just to build the centers.

While they sat idle, the wellness centers occupied valuable real estate inside the stores that could have been put to other, profitable uses. Fed up with waiting, Renda and Bradley put together various ideas for how the space could be utilized. One of them was to staff the centers with nutritionists who would offer dietary advice. Another was to turn them into full-fledged medical clinics run by nurse practitioners. Yet another was to offer telemedicine services. They lobbied Burd to let them implement these plans but, after discussing the matter with Elizabeth, he turned them down. She didn't want to surrender the space, he said.

Behind the scenes, Safeway's board of directors was losing patience. After twenty years in the job, it was clear that Burd had lost the confidence of Wall Street. His first decade as CEO had been a big success and featured a sharp rise in Safeway's stock price. But in recent years, his passion for health and wellness had made him lose sight of what remained at the heart of the company: the unglamorous business of selling groceries. The large investment made in the wellness centers and the endless delays in bringing it to fruition were the last straw.

Shortly after the stock market closed on January 2, 2013, Safeway put out a press release announcing that Burd would retire the following May after the company's annual shareholder meeting. The news was presented as a voluntary decision, but Renda and other executives suspected that the board had asked him to step down. Even on his way out, Burd remained upbeat about the prospects for the still-secret Theranos partnership. Among a list of his achievements as CEO, the press release quoted him as saying that Safeway would soon "be rolling out a wellness initiative that has the potential to transform the Company."

After Burd's departure, the communication

channel to Elizabeth was lost. Anyone from Safeway who wanted to talk to Theranos had to go through Sunny or the Frat Pack. Sunny acted put-off whenever Safeway executives asked for status updates, as if his time was too precious to waste and they had no idea what it took to produce an innovation of this magnitude. His arrogance was infuriating. And yet Safeway was still hesitant to walk away from the partnership. What if the Theranos technology did turn out to be game-changing? It might spend the next decade regretting passing up on it. The fear of missing out was a powerful deterrent.

As for Burd, it was clear he hadn't been ready to retire. Just three months after leaving the supermarket chain, he founded a consulting firm to advise companies on how to reduce their health-care costs. He called it Burd Health. In his new role as a fellow health-startup founder, he tried to get back in touch with Elizabeth. But she no longer returned his calls.

| TEN |

"Who Is LTC Shoemaker?"

Lieutenant Colonel David Shoemaker had been politely listening to the confident young woman seated at the head of the conference table explain how her company intended to operate when, fifteen minutes in, he couldn't hold his tongue anymore.

"Your regulatory structure is not going to fly," he said, interrupting her.

Elizabeth shot an annoyed look at the bespectacled officer in army fatigues as he enumerated the various regulations he thought the approach she'd described fell afoul of. This was not what she wanted to hear. Shoemaker and the little military delegation he was leading had been invited to Palo Alto on this morning in November 2011 to bless Theranos's plans to deploy its devices in the Afghan war theater, not to raise objections about its regulatory strategy.

The idea of using Theranos devices on the battlefield had germinated the previous August when Elizabeth had met James Mattis, head of the U.S. Central Command, at the Marines' Memorial Club in San Francisco. Elizabeth's impromptu pitch about how her novel way of testing blood from just a finger prick could help diagnose and treat wounded soldiers faster, and potentially save lives, had found a receptive audience in the four-star general. Jim "Mad Dog" Mattis was fiercely protective of his troops, which made him one of the most popular commanders in the U.S. military. The hard-charging general was open to pursuing any technology that might keep his men safer as they fought the Taliban in the interminable, atrocity-marred war in Afghanistan. After meeting Elizabeth, he'd asked subordinates at CENTCOM to set up a live field test of the Theranos device.

Under military rules, such requests had to be routed through the army's medical department at Fort Detrick in Maryland, where they usually landed on the desk of Lt. Col. Shoemaker. As deputy director for the Division of Regulated Activities and Compliance, Shoemaker's job was to ensure that the army abided by all laws and regulations when it experimented with medical devices.

Shoemaker wasn't your average military bureaucrat. He had a Ph.D. in microbiology and had spent years doing medical research on vaccines for meningitis and tularemia, a dangerous bacterium found in cottontail rabbits that was weaponized by the United States and the Soviet Union during the Cold War. He'd also been the first army officer to complete a one-year fellowship at the Food and Drug Administration, making him the army's resident expert on FDA regulations.

With his genial smile and his southern Ohio drawl, Shoemaker had a calm, self-effacing manner about him, but he could be direct with people when he needed to be. Theranos's strategy, which envisioned bypassing the FDA altogether, was a nonstarter, he warned Elizabeth, especially if she planned to roll out her devices nationwide by the following spring, as she had asserted to him. There was no way the agency would allow her to do that without going through its review process, he told her.

Elizabeth disagreed forcefully, citing advice Theranos had received from its lawyers. She was so defensive and obstinate that Shoemaker quickly realized that prolonging the argument would be a waste of time. She clearly didn't want to hear anything that contradicted her point of view. As he looked around the table, he noted that she had brought no regulatory affairs expert to the meeting. He suspected the company didn't even employ one. If he was right about that, it was an incredibly naïve way of operating. Health care was the most highly regulated industry in the country and for good reason: the lives of patients were at stake.

Shoemaker told Elizabeth she would need to get something in writing from the FDA supporting her position if she wanted him to greenlight the use of her machines on army personnel. Her face conveyed deep displeasure. She resumed her presentation but gave Shoemaker the cold shoulder for the rest of the day.

IN HIS EIGHTEEN-YEAR CAREER in the army, Shoemaker had come across a lot of people who seemed to think the military was exempt from civilian regulations and free to conduct medical research as it pleased. That was simply not the case, though this wasn't to say it hadn't happened in the past. The Pentagon tested mustard gas on American soldiers during World War II and Agent Orange on prisoners in the 1960s. But the days of unsupervised, freewheeling medical experimentation by the military were long gone.

During the Serbian conflict in the 1990s, for instance, the Pentagon made sure to get the FDA's assent before offering troops deployed in the Balkans an experimental vaccine against tick-borne encephalitis. And only soldiers who wished to receive the vaccine were given it. Similarly, the army worked closely with the agency to make an investigational vaccine against botulinum toxin available to soldiers in Iraq in 2003. At the time, concerns were high that Saddam Hussein had stockpiled the lethal biological agent, and the promising vaccine, which had been developed by researchers at Fort Detrick, hadn't yet been approved by the FDA.

In both instances, the army consulted an institutional review board, or IRB—a committee within the military that monitors medical research to ensure that it is conducted safely and ethically. If the IRB deems that a proposed study doesn't pose significant risks, the FDA will usually allow it to go forward, provided it's carried out under a strict protocol the committee has reviewed and approved.

What was valid for vaccines was also valid for medical devices. If Theranos wanted to try out its blood-testing machines on troops in Afghanistan, Shoemaker felt certain that it would need to put together an IRB-approved study protocol. But since Elizabeth had been so adamant and he was also getting second-guessed by CENTCOM, he decided to bring in Jeremiah Kelly, an army lawyer who'd previously worked at the FDA. He scheduled another meeting with Elizabeth so Kelly could hear from her directly and provide a second opinion. They agreed to meet at 3:30 p.m. on December 9, 2011, at the Washington, D.C., offices of Theranos's law firm, Zuckerman Spaeder.

Elizabeth came to the meeting alone with a single-page document outlining the same regulatory approach Shoemaker had heard her present a few weeks before in Palo Alto. He had to give it to her: the structure she laid out was creative. One might even call it sneaky.

The document explained that Theranos's devices were merely remote sample-processing units. The real work of blood analysis would take place in the company's lab in Palo Alto, where computers would analyze the data the devices transmitted to it and qualified laboratory personnel would review and interpret the results. Hence only the Palo Alto lab needed to be certified. The devices themselves were akin to "dumb" fax machines and exempt from regulatory oversight.

There was a second wrinkle Shoemaker found equally hard to swallow: Theranos maintained that the blood tests its devices performed were laboratory-developed tests and therefore beyond the FDA's purview.

The Theranos position then was that a CLIA certificate for its Palo Alto lab was sufficient for it to deploy and use its devices anywhere. This was a clever theory, but Shoemaker didn't buy it. And neither did Kelly. The Theranos devices were more than just dumb fax machines. They were blood analyzers and, like all other blood analyzers on the market, they would eventually need to be reviewed and approved by the FDA. Until then, Theranos would need to consult with an institutional review board and come up with a study protocol that the agency could live with. It was a process that typically took six to nine months.

Elizabeth continued to disagree despite the army lawyer's presence. Her body language wasn't as hostile as it had been in Palo Alto and she was more willing to engage in a discussion, but they remained at an impasse. What was strange was that no one from Zuckerman Spaeder was there with her in the room. Shoemaker had expected her to show up accompanied by several of the firm's partners, but there she was on her own. She continued to invoke the firm's legal advice, but no one from the firm was present to attest to it.

The meeting ended with Shoemaker reiterating that he would need to see something in writing from the FDA backing up Theranos's regulatory stance before he signed off on any experiment in Afghanistan. Elizabeth agreed to get such a letter. She acted as if it was a formality. Shoemaker very much doubted so, but at least things were now clear: the ball was in Theranos's court.

SHOEMAKER DIDN'T HEAR anything more about the matter until the late spring of 2012, when he started receiving queries from CENTCOM again. He couldn't help but be annoyed. Not only had Theranos failed to produce the letter he'd asked for, the company had gone completely silent since he and Kelly had traveled to Washington to meet Elizabeth in December.

With the approval of his boss, he decided to make contact with the FDA himself. On the morning of June 14, 2012, he sent an email to Sally Hojvat, the head of the agency's microbiology devices division. The two had worked together during Shoemaker's FDA fellowship in 2003 and had just run into each other at a conference the previous week. Shoemaker described to Hojvat the Theranos situation and, calling the company's regulatory approach "quite novel," requested the agency's guidance on it. Although he didn't intend his email as anything more than an informal request for advice, it set off a sequence of events that would have made him think twice about sending it if he could have foreseen them.

Hojvat forwarded his query to five of her colleagues, including Alberto Gutierrez, the director of the FDA's Office of In Vitro Diagnostics and Radiological Health. Gutierrez, who had a Ph.D. in chemistry from Princeton, happened to have spent a not insignificant portion of his twenty-year career at the agency pondering the question of laboratory-developed tests.

The FDA had long considered it within its power to regulate LDTs, as laboratory-developed tests were known. However, in practice, it had not done so because back in 1976, when the Federal Food, Drug, and Cosmetic Act was amended to expand the agency's authority from drugs to medical devices, LDTs weren't common. They were only made by local laboratories occasionally when an unusual medical case required it.

That changed in the 1990s when laboratories started to make more complex tests for mass use, including genetic tests. By the FDA's own reckoning, scores of flawed and unreliable tests had since been marketed for conditions ranging from whooping cough and Lyme disease to various types of cancers, resulting in untold harm to patients. There was a growing consensus within the agency that it needed to start policing this part of the lab business, and the biggest proponent of that view was Gutierrez. When he saw the email Hojvat forwarded to him from Shoemaker, Gutierrez shook his head in disbelief. The approach it described was exactly the type of regulatory end run around the FDA that he wanted to put a stop to.

Gutierrez's view that it was the FDA, not the Centers for Medicare and Medicaid Services, that should regulate LDTs did not mean he didn't get along with his colleagues at CMS. To the contrary, they had a good working relationship and often communicated across agency lines to try to bridge the regulatory gap spawned by outdated statutes. Gutierrez forwarded the Shoemaker email to Judith Yost and Penny Keller, two members of CMS's lab-oversight division, adding a note at the top:

> How about this one!!! Would CMS consider this an LDT? I have a hard time seeing that we would exercise enforcement discretion on this one.
>
> Alberto

After some back-and-forth, Gutierrez, Yost, and Keller all reached the same conclusion: the Theranos model didn't comply with federal regulations. Yost and Keller decided it wouldn't hurt to send someone to Palo Alto to see what exactly this company that none of them had heard of before was up to and to correct its misconceptions.

The job fell to Gary Yamamoto, a veteran field inspector in CMS's regional office in San Francisco. Two months later, on August 13, 2012, Yamamoto arrived unannounced at Theranos's offices in Palo Alto. By then, the company had completed its move to the old Facebook building located at 1601 South California Avenue, less than a mile from its former home on Hillview Avenue.

Sunny and Elizabeth ushered Yamamoto into a conference room. When he explained that his agency had received a complaint about Theranos and that he was there to look into it, he was surprised to learn that they knew where and who it came from. Someone had apparently tipped them off about Shoemaker's email to the FDA. Elizabeth was not pleased, a sentiment made clear by the scowl on her face. She and Sunny professed not to know what Shoemaker had been talking about in his email. Yes, Elizabeth had met with the army officer, but she had never told him Theranos intended to deploy its blood-testing machines far and wide under the cover of a single CLIA certificate.

Why then **had** Theranos applied for a CLIA certificate? Yamamoto asked. Sunny responded that the company wanted to learn about how labs worked and what better way to do that than to operate one itself? Yamamoto found that answer fishy and borderline nonsensical. He asked to see their lab.

They couldn't deny him access the way they had to Kevin Hunter. This was the representative of a federal regulatory agency, not some private lab consultant they could thumb their noses at. So Sunny reluctantly took the inspector to a room on the second floor of the new building. After Dupuy's firing, Theranos had moved the lab there from its temporary East Meadow Circle location.

What Yamamoto found in the room didn't impress him but didn't raise any big concerns either: it was a small space with a couple of people in white lab coats and a smattering of commercial diagnostic instruments that were sitting idle. It looked like any other lab. No sign of any special or novel blood-testing technology. When he pointed this out, Sunny said the Theranos devices were still under development and the company had no plans to deploy them without FDA clearance—flatly contradicting what Elizabeth had told Shoemaker on not one but two occasions. Yamamoto wasn't sure what to believe. Why would the army officer have made all that stuff up?

There were no clear violations he could point to about the way Theranos was currently operating, however, so he let Sunny off with a long lecture about lab regulations. He made sure to emphasize that the scenario Shoemaker had described in his email to Sally Hojvat—experimental blood analyzers operated remotely from one CLIA-certified mother base—was out of the question. If Theranos intended to eventually roll its devices out to other locations, those places would need CLIA certificates too. Either that or, better yet, the devices themselves would need to be approved by the FDA.

ELIZABETH WASN'T ONE to sit still and quietly take it when she felt her company was under attack. In a blistering email to General Mattis, she hit back against the person who had dared to put hurdles in her way. Shoemaker, she wrote, had communicated "blatantly false information" to the FDA and CMS about Theranos. She went on to heap several paragraphs of scorn on the lieutenant colonel and listed seven inaccurate statements he had allegedly made to the agencies "compiled with assistance from our counsel." Her email closed with a request:

> We are taking swift action to correct these misleading statements. I would very much appreciate your help in getting this information corrected with the regulatory agencies—LTC Shoemaker communicated to the FDA that he was giving them "a heads-up" about "what Theranos was up to" and provided the agency with incorrect information that makes us appear to be in violation of the law. Since this misinformation came from within DoD, it will be invaluable if this information is formally corrected by the right people in DoD. Thank you for your thoughts and as always for your time.
>
> With my best regards,
> Elizabeth.

When he read Elizabeth's email a few hours later, Mattis was furious. He forwarded it to Colonel Erin Edgar, CENTCOM's command surgeon and the aide-de-camp he'd put in charge of making the Theranos field test happen, with a note that conveyed his anger:

> Erin: Who is LTC Shoemaker and what is going on here? . . . I have tried to get this device tested in theater asap, legally and ethically, and I need to know did this visit happen as related below and how do we overcome this new obstacle . . . Bottom line, I need ground truth for the

> accuracy of the statements below. If I need to see LTC Shoemaker and LTC Mann so they can explain how I am pushing for something unethical or illegal please set up a time for them to meet with me in Tampa when I return to the states [**sic**] (I am going to be delayed in theater past my original return date). Thanks, M

The CMS inspector's surprise visit had put Elizabeth on the warpath. In a phone call to Colonel Edgar, she threatened to sue Shoemaker. Edgar relayed her threat to his Fort Detrick colleague, along with news of the inspection. He also forwarded to Shoemaker Elizabeth's email to Mattis and Mattis's reaction to it.

When he read the email string, Shoemaker blanched. Mattis was one of the most powerful and fearsome people in the military. The blunt-spoken general had once famously told Marines stationed in Iraq, "Be polite, be professional, but have a plan to kill everybody you meet." This was not a guy you wanted to get on the wrong side of if you were a lower-ranking army officer.

Shoemaker also felt genuinely bad that his actions had led to an inspection of the company. He was well placed to know how unpleasant such visits could be: his previous assignment had been at the Army Medical Research Institute of Infectious Diseases, where he'd taken over as director of biosurety, the department responsible for securing biothreat agents used in army research, two weeks before Bruce Ivins killed himself in July 2008. The suicide had led to the disclosure that Ivins, an institute researcher, was the likely perpetrator of the 2001 anthrax attacks and to an avalanche of inspections from an alphabet soup of government agencies that had continued unabated for two years. Shoemaker had been the officer on the receiving end of every single one of them.

With Colonel Edgar's encouragement, he tried to defuse the situation by emailing CMS officials that he had never meant to imply that Theranos had already implemented the regulatory strategy he'd described, merely that it was considering it. He also expressed surprise that the agency had told Theranos he was the one who requested the inspection. The response he got brought another surprise: CMS had told Theranos no such thing; the company already had a copy of his correspondence with the FDA when the inspector arrived.

When he confronted Colonel Edgar with that information, Edgar sheepishly admitted that he had been the one who'd shared his email to Sally Hojvat with Elizabeth in what he described as an oversight. He apologized and invited Shoemaker to come to CENTCOM headquarters in Tampa, Florida, the following week to walk Mattis through the regulatory issues. Shoemaker was nervous about meeting the general face-to-face, but he accepted the invitation. He reached out to Alberto Gutierrez to see if he could join him on the trip, figuring his opinion would carry more weight if it was supported by someone high up at the FDA. Although it was on short notice, Gutierrez agreed to come along.

AT 3:00 P.M. SHARP on August 23, 2012, Colonel Edgar escorted the two men into Mattis's office on MacDill Air Force Base in Tampa. The sixty-one-year-old general was an intimidating figure in person: muscular and broad shouldered, with dark circles under his eyes that suggested a man who didn't bother much with sleep. His office was decorated with the mementos of a long military career. Amid the flags, plaques, and coins, Shoemaker's eyes rested briefly on a set of magnificent swords displayed in a glass cabinet. As they sat down in a wood-paneled conference room off to one side of the office, Mattis cut to the chase: "Guys, I've been trying to get this thing deployed for a year now. What's going on?"

Shoemaker had gone over everything again with Gutierrez and felt confident he was on solid ground. He spoke first, giving a brief overview of the issues raised by an in-theater test of the Theranos technology. Gutierrez took over from there and told the general his army colleague was correct in his interpretation of the law: the Theranos device was very much subject to regulation by the FDA. And since the agency hadn't yet reviewed and approved it for commercial use, it could only be tested on human subjects under strict conditions set by an institutional review board. One of those conditions was that the test subjects give their informed consent—something that was notoriously hard to obtain in a war zone.

Mattis was reluctant to give up. He wanted to know if they could suggest a way forward. As he'd put it to Elizabeth in an email a few months earlier, he was convinced her invention would be "a game-changer" for his men. Gutierrez and Shoemaker proposed a solution: a "limited objective experiment" using leftover de-identified blood samples from soldiers. It would obviate the need to obtain informed consent and it was the only type of study that could be put together as quickly as Mattis seemed to want to proceed. They agreed to pursue that course of action. Fifteen minutes after they'd walked in, Shoemaker and Gutierrez shook Mattis's hand and walked out. Shoemaker was immensely relieved. All in all, Mattis had been gruff but reasonable and a workable compromise had been reached.

The limited experiment agreed upon fell short of the more ambitious live field trial Mattis had had in mind. Theranos's blood tests would not be used to inform the treatment of wounded soldiers. They would only be performed on leftover samples after the fact to see if their results matched the army's regular testing methods. But it was something.

Earlier in his career, Shoemaker had spent five years overseeing the development of diagnostic tests for biological threat agents and he would have given his left arm to get access to anonymized samples from service members in theater. The data generated from such testing could be very useful in supporting applications to the FDA.

Yet, over the ensuing months, Theranos inexplicably failed to take advantage of the opportunity it was given. When General Mattis retired from the military in March 2013, the study using leftover de-identified samples hadn't begun. When Colonel Edgar took on a new assignment as commander of the Army Medical Research Institute of Infectious Diseases a few months later, it still hadn't started. Theranos just couldn't seem to get its act together.

In July 2013, Lieutenant Colonel Shoemaker retired from the army. At his farewell ceremony, his Fort Detrick colleagues presented him with a "certificate of survival" for having the courage to stand up to Mattis in person and emerging from the encounter alive. They also gave him a T-shirt with the question, "What do you do after surviving a briefing with a 4 star?" written on the front. The answer could be found on the back: "Retire and sail off into the sunset."

| ELEVEN |

Lighting a Fuisz

The doorbell at 1238 Coldwater Canyon Drive in Beverly Hills rang at 10:15 a.m. on Saturday, October 29, 2011. The gated one-story Italian villa shrouded by palm trees belonged to Richard and Lorraine Fuisz. The couple had purchased it two years earlier to live closer to their children, who had both moved from Washington, D.C., to Los Angeles after graduating from Georgetown University.

When Richard Fuisz opened the door, a process server tried to hand him a stack of legal papers.

"I'm here to serve a lawsuit on Fuisz Technologies," the man said.

Fuisz told him he couldn't accept service because the company, though it bore his name, was no longer his. He had sold it more than a decade earlier. It was now part of Canadian drugmaker Valeant Pharmaceuticals, he explained. The man placed a call and repeated what Fuisz had said. The response, conveyed by someone shouting on the other end of the line, was that he was at the right address and to just serve the papers. But Fuisz continued to refuse to take them. Losing his patience, the process server threw them at his feet and left. Fuisz took out his cell phone and snapped a photo of the pile scattered across the sidewalk. He knew very well what this was about. As a codefendant in the lawsuit, he'd been personally served with a similar set of papers two days earlier. After mulling things over for a minute, he crouched down and picked up the mess. He decided he didn't want the neighbors to see it.

The lawsuit had been filed by Theranos in federal court in San Francisco. It alleged that he had conspired with Joe and John Fuisz, his sons from his first marriage, to steal confidential patent information from the company and used it to file his own rival patent. The theft, the suit alleged, had been carried out by John on behalf of his father while he was employed at Theranos's former patent counsel, McDermott Will & Emery.

The top of the complaint's first page showed that Theranos had hired the famous lawyer David Boies to represent it. However, as renowned as Boies was, someone at his firm had fumbled their research and named the wrong Fuisz company. It was Fuisz Pharma, Richard and Joe's new firm—not Fuisz Technologies—that the patent in question had been assigned to. Fuisz had refused service because he wanted to make Boies sweat his mistake.

Fuisz and his sons were angered by the suit, but they weren't overly worried about it at first. They were confident in the knowledge that its allegations were false. The first and only time Fuisz had brought up Elizabeth Holmes's startup with John was in an email he had sent his son in July 2006 with a link to a Theranos patent application he'd spotted in the patent office's public database. The email, which was sent more than two months **after** Fuisz filed his own provisional patent, asked if John knew who at McDermott had worked on the Theranos application. John replied that McDermott was a large firm and that he had no idea. The exchange had scarcely registered with John. More than five years later, he had no memory of it. As far as he was concerned, this lawsuit was the first time he'd seen or heard the word "Theranos."

John had no reason to wish Elizabeth or her family ill; on the contrary. When he was in his early twenties, Chris Holmes had written him a letter of recommendation that helped him gain admission to Catholic University's law school. Later, John's first wife had gotten to know Noel Holmes through Lorraine Fuisz and become friendly with her. Noel had even dropped by their house when John's first son was born to bring the baby a gift.

Moreover, Richard and John Fuisz weren't close. John thought his father was an overbearing megalomaniac and tried to keep their interactions to a bare minimum. In 2004, he'd even dropped him as a McDermott client because he was being difficult and slow to pay his bills. The notion that John had willingly jeopardized his legal career to steal information for his father betrayed a fundamental misunderstanding of their frosty relationship.

But Elizabeth was understandably furious at Richard Fuisz. The patent application he had filed in April 2006 had matured into U.S. Patent 7,824,612 in November 2010 and now stood in the way of her vision of putting the Theranos device in people's homes. If that vision was someday realized, she would have to license the bar code mechanism Fuisz had thought up to alert doctors to patients' abnormal blood-test results. Fuisz had rubbed that fact in her face the day his patent was issued by sending a Fuisz Pharma press release to info@theranos.com, the email address the company provided on its website for general queries. Rather than give in to what she saw as blackmail, Elizabeth had decided to steamroll her old neighbor by hiring one of the country's best and most feared attorneys to go after him.

DAVID BOIES'S LEGEND preceded him. He had risen to national prominence in the 1990s when the Justice Department hired him to handle its antitrust suit against Microsoft. On his way to a resounding courtroom victory, Boies had grilled Bill Gates for twenty hours in a videotaped deposition that proved devastating to the software giant's defense. He had gone on to represent Al Gore before the Supreme Court during the contested 2000 presidential election, cementing his status as a legal celebrity. More recently, he'd successfully led the charge to overturn Proposition 8, California's ban on gay marriage.

Boies was a master litigator who could be ruthless when he felt the situation required it. In one case that illustrated his no-holds-barred ethos, he had escalated a business dispute between a client and the owner of a small Palm Beach lawn-care company into a federal racketeering lawsuit in which he'd accused the man and three of his gardeners of conspiracy, fraud, extortion, and—last but not least—violating antitrust law. After a judge in Miami dismissed the suit, Boies appealed the verdict to the U.S. Court of Appeals for the Eleventh Circuit in Atlanta. Only after that appeal failed did he drop the case.

Boies's law firm, Boies, Schiller & Flexner, had gained a reputation for aggressive tactics. It didn't take the Fuiszes long to find out why. In the weeks before Theranos sued, all three had picked up clues that they were under surveillance. Richard Fuisz noticed a car tailing him when he drove to Van Nuys Airport to board a flight to Las Vegas. Joe, who lived in Miami, was warned by his neighbor, a retired cop who acted like his block's self-appointed captain, that someone was watching his house. John and his wife spotted a man taking pictures of their home in Georgetown. The Fuiszes now felt sure those had been private investigators hired by Boies.

The surveillance continued after the suit was filed and unnerved Fuisz's wife, Lorraine. Cars were frequently parked across the street from their Beverly Hills house, a driver sitting idle inside. One day, Lorraine noticed that the person behind the wheel was a blond woman and became convinced that it was her old friend, Noel Holmes. Fuisz thought that was unlikely but grabbed his camera and a telephoto lens and took a picture of the car, a gray Toyota Camry, from inside the house. He then walked outside to confront the driver. As he approached the car, it sped off. When he later took a closer look at the photo, he couldn't make out the woman's face well enough to rule Noel out. That upset Lorraine even more. She felt sure the Holmeses were out to bankrupt them and take possession of their house. She became nearly hysterical.

Boies's use of private investigators wasn't just an intimidation tactic, it was the product of a singular paranoia that shaped Elizabeth and Sunny's view of the world. That paranoia centered on the belief that the lab industry's two dominant players, Quest Diagnostics and Laboratory Corporation of America, would stop at nothing to undermine Theranos and its technology. When Boies had first been approached about representing Theranos by Larry Ellison and another investor, it was that overarching concern that had been communicated to him. In other words, Boies's assignment wasn't just to sue Fuisz, it was to investigate whether he was in league with Quest and LabCorp. The reality was that Theranos was on neither company's radar at that stage and that, as colorful and filled with intrigue as Fuisz's life had been, he had no connection to them whatsoever.

Two months after Theranos filed its lawsuit, Keker & Van Nest, the law firm John Fuisz hired to defend him, sent Boies several documents that went a long way toward disproving Theranos's accusations. One of them was a declaration by Brian McCauley, McDermott's records manager, stating that a thorough search of the firm's records management and email systems had shown that neither John nor his secretary had ever accessed any Theranos files. Attached to the declaration were exhibits showing all the steps McCauley had taken to arrive at his conclusion. But in a response five days later, Boies dismissed the documents as "self serving" and "not . . . very persuasive."

Richard Fuisz tried to appeal directly to the

Theranos board by sending its members several letters. In one of them, he included a photo of Elizabeth as a child to make the point that the families had once been on friendly terms and had known each other a long time. With another, he enclosed a binder filled with copies of all the emails he and his patent attorney had exchanged leading up to his April 2006 patent application to show that the patent stemmed from his own work. He also offered to meet with the board members. The only response he got was from Boies, who wrote back that Theranos was "puzzled" as to why he thought the emails proved anything.

ALTHOUGH HE DIDN'T have a shred of evidence to prove that John Fuisz had done what Theranos alleged, there were some things in John's past that Boies planned to use to sow doubts in the minds of a judge or jury.

In 1992, when John was fresh out of law school, he had acted as a courier between his father and a college friend who worked at the law firm Skadden, Arps, Slate, Meagher & Flom. The college friend had given John a stack of Skadden billing documents to pass on to his dad. At the time, Richard Fuisz was in a legal battle with a Skadden client, the heavy-equipment maker Terex Corporation, which had sued him for libel for telling a congressional committee that it sold Scud missile launchers to Iraq. Even though the incident was twenty years old and the libel suit had been settled without any court finding that John had done anything wrong, Boies intended to use it to suggest he had a history of funneling stolen information to his father.

There was something else Boies planned to exploit that was both more recent and potentially more damaging: McDermott had forced John to resign in 2009 after he clashed with the firm's brass in an unrelated matter. The cause of the dispute had been John's insistence that the firm withdraw its reliance on a forged document in a case before the International Trade Commission in which McDermott was representing a Chinese state-owned company against the U.S. government's Office of Unfair Import Investigations. McDermott's leadership had agreed to withdraw the document, but the move had severely weakened the Chinese client's defense and angered the firm's senior partners. The firm had subsequently asked John to leave, citing a list of other incidents that it said showed a pattern of behavior unbecoming of a partner. One of the reasons cited was a complaint a client had made about John. At the time, the firm had refused to tell John who the client was or what the complaint was about, but he now figured it must have been Elizabeth's September 2008 complaint to Chuck Work about his father's patent.

Boies's strategy of painting John Fuisz in a bad light was dealt a setback in June 2012 when the judge overseeing the case dismissed all the claims against John on the grounds that California's one-year statute of limitations for legal malpractice had expired. Boies turned around and sued McDermott in state court in Washington, D.C., but that suit was soon dismissed when the court ruled that Theranos's allegations against John and the firm were completely speculative. "Simply because attorneys within the firm had access [to the Theranos documents] does not mean the firm did not maintain confidentiality," the judge wrote.

Yet Boies was still in business: while dismissing the claims against John, the judge in the California case had allowed many of the claims against Richard and Joe Fuisz to stand and the case to proceed to trial. John might no longer be a defendant, but Boies could still use the same narrative of collusion between father and son to argue his case against Richard and Joe.

As the litigation dragged on into the fall, John's initial annoyance with the case morphed into full-blown fury at Elizabeth. After leaving McDermott, he had founded his own practice and the Theranos case and its allegations had cost him several clients. Opposing counsel had also brought them up to tar him in two of his cases. By the time Boies Schiller attorneys deposed John in the spring of 2013, his anger was made rawer by another source of stress: his wife, Amanda, had been diagnosed with a vasa previa, a complication of pregnancy in which the fetus's blood vessels are exposed and at risk of rupture. She and John were in anxious limbo until the baby reached thirty-four weeks and doctors could deliver her and place her in the neonatal intensive care unit.

Even in the best of times, John had a short fuse. Growing up, he often got into fights with other boys. Under the questioning of one of Boies's partners that day, he became combative and ornery, using foul language and letting his temper flare. At the end of the six-and-a-half-hour deposition, he made a threat that played right into Boies's hands. Asked by one of his father's lawyers whether the case had caused him reputational harm and, if so, whether that had affected his demeanor during the deposition, he responded:

> Absolutely, I am more than pissed off at these people. I intend to seek my revenge and sue the fuck out of them when this is over, and you can guarantee I will not let Elizabeth Holmes have another fucking company as long as she lives. I will use my ability to file patents and fuck with her till she dies, absolutely.

While John Fuisz's anger was boiling over, his father and brother were becoming concerned with how expensive the litigation was getting. They had hired the Los Angeles law firm Kendall Brill & Klieger to represent them at a cost of about $150,000 a month. Laura Brill, the partner who was handling their case, wanted to file an anti-SLAPP motion to try to get the Theranos suit thrown out as frivolous, which would have cost an additional $500,000 with little assurance that it would succeed. They decided to switch to a smaller and less expensive Northern California firm called Banie & Ishimoto and hired the George Washington University Law School professor Stephen Saltzburg, who had done legal work for Fuisz in the past, to oversee its work.

On the other side, they knew they were up against one of the most expensive lawyers in the world. Boies charged clients nearly a thousand dollars an hour and was reputed to earn more than $10 million a year. What they didn't know, however, was that in this instance he had accepted stock in lieu of his regular fees. Elizabeth had granted his firm 300,000 Theranos shares at a price of $15 a share, which put a sticker price of $4.5 million on Boies's services.

It wasn't the first time Boies had worked out an alternative fee arrangement with a client and taken stock for payment. During the dot-com boom, he had accepted shares to represent WebMD, the website that provides medical information to consumers. Boies had a venture capitalist's approach to cases and figured he and his firm stood to earn a lot more money by getting paid in stock. But it also meant he had a vested financial interest in Theranos that made him more than just its legal advocate. It helped explain why, in early 2013, Boies began attending all of the company's board meetings.

EVEN THOUGH Elizabeth's name was on all of Theranos's patents, Richard Fuisz was highly skeptical that a college dropout with no medical or scientific training had done much real inventing. What was more likely, he thought, was that other employees with advanced degrees had done the work she'd patented.

As the two sides prepared for trial, Fuisz noticed one name that appeared as a co-inventor on many of Elizabeth's patents: Ian Gibbons. With a little research, he learned a few basic facts about the man. Gibbons was a Brit who had a Ph.D. in biochemistry from the University of Cambridge, and he was credited as an inventor on some fifty U.S. patents, including nineteen stemming from his work at a company called Biotrack Laboratories in the 1980s and 1990s.

Fuisz presumed that Gibbons was a legitimate scientist and that, like most scientists, he was an honest person. If he could get him to admit under oath that there was nothing in his patent that borrowed from, or was similar to, Elizabeth's early patent applications, it would deal a big blow to Theranos's case. He and Joe had also noticed that some of Gibbons's Biotrack patents were similar to the Theranos ones, opening the company up to charges that it had improperly recycled some of his past work. They added Gibbons's name to the list of witnesses they wanted to depose. But then something strange happened: over the next five weeks, the Boies Schiller attorneys kept ignoring their request to schedule Gibbons's deposition. Suspicious, the Fuiszes asked their lawyers to press the matter.

| TWELVE |

Ian Gibbons

Ian Gibbons was the first experienced scientist Elizabeth had hired after launching Theranos. He came recommended by her Stanford mentor, Channing Robertson. Ian and Robertson had met at Biotrack in the 1980s, where they had invented and patented a new mechanism to dilute and mix liquid samples.

From 2005 to 2010, Ian led Theranos's chemistry work alongside Gary Frenzel. Ian, who had joined the startup first, was initially senior to Gary. But Elizabeth soon inverted their roles because Gary had better people skills, which made him a smoother manager. The two of them cut quite a contrast—Ian, the reserved Englishman with a wry sense of humor, and Gary, the garrulous former rodeo rider who spoke with a Texas twang. But they had a good relationship grounded in their respect for each other as scientists and would sometimes roast each other in meetings.

Ian fit the stereotype of the nerdy scientist to a T. He wore a beard and glasses and hiked his pants high above his waist. He could spend hours on end analyzing data and took copious notes documenting everything he did at work. This meticulousness carried over to his leisure time: he was an avid reader and kept a list of every single book he'd read. It included Marcel Proust's seven-volume opus, **Remembrance of Things Past,** which he reread more than once.

Ian and his wife, Rochelle, had met in the early 1970s at Berkeley. He had come over from England to do a postdoctorate fellowship in the university's department of molecular biology, where Rochelle was doing her graduate research. They'd never had children, but Ian doted on their dogs Chloe and Lucy and on Livia, a cat he'd named after the wife of the Roman emperor Augustus.

Besides reading, Ian's other two hobbies were going to the opera— he and Rochelle regularly went to San Francisco's War Memorial Opera House and in the summer flew to New Mexico to attend open-air performances of the Santa Fe Opera at dusk—and photography. For laughs, he liked altering photos. One of the many he doctored showed him as a gloved and bow-tied mad scientist mixing blue and purple potions. In another, he inserted himself into the foreground of a portrait of the British royal family.

As a biochemist, Ian's specialty was immunoassays, which was the main reason Theranos had focused its early efforts on that class of test. He was passionate about the science of blood testing and loved to teach it. In the company's early years, he would sometimes hold little lectures to educate the rest of the staff about the fundamentals of biochemistry. He also did presentations about how to create various blood tests that were recorded and stored on the company's servers.

One source of recurring tension between Ian and the Theranos engineers was his insistence that the blood tests that he and the other chemists designed perform as accurately inside the Theranos devices as they did on the lab bench. The data he collected suggested that was rarely the case, which caused him considerable frustration. He and Tony Nugent butted heads over this issue during the development of the Edison. As admirable as Ian's exacting standards were, Tony felt that all he did was complain and that he never offered any solutions.

Ian also had issues with Elizabeth's management, especially the way she siloed the groups off from one another and discouraged them from communicating. The reason she and Sunny invoked for this way of operating was that Theranos was "in stealth mode," but it made no sense to Ian. At the other diagnostics companies where he had worked, there had always been cross-functional teams with representatives from the chemistry, engineering, manufacturing, quality control, and regulatory departments working toward a common objective. That was how you got everyone on the same page, solved problems, and met deadlines.

Elizabeth's loose relationship with the truth was another point of contention. Ian had heard her tell outright lies more than once and, after five years of working with her, he no longer trusted anything she said, especially when she made representations to employees or outsiders about the readiness of the company's technology.

Ian's frustrations bubbled over in the fall of 2010 as Theranos's courtship of Walgreens intensified. He complained to his old friend Channing Robertson. Ian thought Robertson would keep their conversation private, but he reported everything Ian had said to Elizabeth. Rochelle was in bed when Ian arrived at their Portola Valley home late that Friday night. He told his wife that Robertson had betrayed his confidence and that Elizabeth had fired him.

To their surprise, Sunny called the next day. Unbeknownst to Ian, in the intervening hours several of his colleagues had lobbied Elizabeth to reconsider. Sunny offered Ian his job back, albeit without the same responsibilities. Ian had been head of the general chemistry group, which was in charge of creating new blood tests beyond the immunoassays they'd developed for the Edison, when Elizabeth fired him. He was allowed to come back as a technical consultant to the group, but its leadership was given to Paul Patel, a biochemist who had been hired two months earlier on Ian's recommendation.

Ian was a proud man and he took the demotion hard. The humiliation he felt was compounded when, eighteen months later, the company moved to the old Facebook building and he lost the private office he'd had at the Hillview Avenue headquarters. To be sure, he wasn't the only one being marginalized by then: Gary Frenzel and Tony Nugent too were being sidelined as Elizabeth and Sunny hired and promoted newer recruits over them. It was as if the company's old guard—the people who had gotten Elizabeth to this point—was being mothballed.

A FEW MONTHS BEFORE the move, Tony had noticed a poster for the movie **Women in Love** in Ian's office and they got to talking. The 1969 film was based on D. H. Lawrence's novel of the same name about the relationships between two sisters and two men in an English mining town around the time of the First World War. Ian mentioned that he'd toured Ireland when it came out, which would have coincided with when Tony was still a child growing up there. That led to other musings. Tony learned that Ian's father had been captured in

North Africa during World War II. After initially being detained in a POW camp in Italy, he had been marched across Europe to a different camp in Poland, where he was liberated at the end of the war.

The conversation eventually drifted back to the here and now and to Theranos. Tony, who like Ian no longer had Elizabeth's favor and was being excluded from the development of the miniLab, floated the notion that perhaps the company was just a vehicle for Elizabeth and Sunny's romance and that none of the work they did really mattered.

Ian nodded. "It's a folie à deux," he said.

Tony didn't know any French, so he left to go look up the expression in the dictionary. The definition he found struck him as apt: "The presence of the same or similar delusional ideas in two persons closely associated with one another."

After the move to the old Facebook building, Ian grew more sullen. He was relegated to a desk in the general population of employees with his back facing a wall. It was a symbol of how unimportant he'd become.

One day, the engineer Tom Brumett ran into him at Fish Market, a seafood restaurant on El Camino Real where he was meeting a friend. As they stood in line waiting for a table, Ian asked if he could join them. Tom and Ian were both in their mid-sixties and had established a friendly rapport. The first time they'd interacted was shortly after Tom came to work at Theranos in 2010. Upset that Sunny and other managers were disregarding his opinion during a discussion about what sort of engineering personnel should be hired to assist him, Tom had walked out of the meeting in a huff with thoughts of quitting. Ian had come running after him and assured him that his opinion did matter—a gesture Tom had greatly appreciated.

Over the next two years, Tom had noticed Ian's growing gloom. As they sat down for lunch at Fish Market, Tom wondered whether Ian had followed him there. Most Theranos employees ate the food Elizabeth and Sunny had catered and didn't leave the office during the day. What's more, the restaurant wasn't near the office and Ian had walked in just a minute or two after him. Ian had probably hoped to catch him alone, Tom thought. He seemed desperate for someone to talk to. But Tom was there to reconnect with his friend, a salesman for a Japanese chipmaker. They tried to include Ian in the conversation, but he remained quiet after an initial exchange of pleasantries. Later, when he replayed the scene in his mind, Tom realized he'd ignored his colleague's silent cry for help.

Tom ran into Ian one last time in early 2013 in the office cafeteria. By then, he looked despondent. Tom tried to buck him up, reminding him that he was earning decent money and urging him not to take his work predicament so seriously. It was just a job, after all. But Ian just stared at his plate, disconsolate.

IAN'S DEMOTION WASN'T the only thing eating away at him. Although he was now a mere in-house consultant, he continued to work closely with the person who had taken his job, Paul Patel. Paul had tremendous respect for Ian as a scientist. When he was in graduate school in England, he had read all about the pioneering work on immunoassays Ian had done in the 1980s at a company called Syva.

After he was promoted, Paul continued to treat Ian as an equal and to consult with him about everything. But they differed in one crucial respect: Paul shied away from conflict and was more willing to compromise with the engineers building the miniLab than Ian was. Ian refused to give an inch and became furious when he felt he was being asked to lower his standards. Paul spent numerous evenings on the phone with him trying to calm him down. During these discussions, Ian told Paul to stand by his convictions and never to lose sight of his concern for the patient.

"Paul, it has to be done right," Ian would say.

Sunny had put a man named Samartha Anekal, who had a Ph.D. in chemical engineering but no industry experience, in charge of integrating the various parts of the miniLab. Sam was perceived by some of his colleagues as a yes-man who did Sunny's bidding. Throughout 2012, Ian and Paul had several tense meetings with Sam. Ian stormed out of one of them after Sam informed them that the miniLab's spectrophotometer didn't yet meet certain specifications Ian considered nonnegotiable. Sam had previously agreed to them but now said he needed more time. When he returned to his desk, Ian was distraught.

On weekends, Ian and Rochelle often went on walks in the rolling hills surrounding Portola Valley with Chloe and Lucy, their two American Eskimo dogs. During one of these walks, Ian told Rochelle that nothing at Theranos was working, but he didn't go into any details. The strict nondisclosure agreements he was bound by prevented him from discussing anything specific about the company, even with his wife. He also bemoaned the turn his career had taken. He felt like an old piece of furniture that had been warehoused. Elizabeth and Sunny had long stopped listening to him.

In the early months of 2013, Ian stopped going into the office on most days and instead worked from home. He'd been diagnosed with colon cancer six years earlier and had missed some time at work after undergoing surgery and chemotherapy. Colleagues assumed that the cancer had returned. But that was not the case. He remained in remission and his physical health was fine. The problem lay with his mental health: he was in the throes of a deep, undiagnosed clinical depression.

IN APRIL, Theranos informed Ian that he had been subpoenaed to testify in the Fuisz case. The prospect of being deposed made him nervous. He and Rochelle discussed the lawsuit several times. Rochelle had once done work as a patent attorney, so Ian asked her to review Theranos's patent portfolio in the hope that she could give him some advice. While doing so, she noticed that Elizabeth's name was on all the company's patents, often in first place in the list of inventors. When Ian told her that Elizabeth's scientific contribution had been negligible, Rochelle warned him that the patents could be invalidated if this was ever exposed. That only served to make him more agitated.

Ian couldn't tell whether there was any foundation for Elizabeth's theft allegation when he read the Fuisz patent and the early Theranos patent applications side by side. But he knew one thing for sure: he didn't want to be involved in the case. And yet he worried that his job depended on it. He'd started drinking heavily in the evenings. He told Rochelle that he didn't think he could ever resume a normal schedule at Theranos. The thought of going back to the office made him sick, he said. Rochelle told him he should quit if the job made him that miserable. But resigning didn't seem like an option to him. At age sixty-seven, he didn't think he would be able to find another job. He also clung to the idea that he could still help the company fix its problems.

On May 15, Ian contacted Elizabeth's assistant to schedule a meeting with her, hoping to work out some sort of alternative employment arrangement. But when the assistant called back to confirm a meeting for the next day, Ian became anxious. He told Rochelle he was worried that Elizabeth would use the meeting to fire him. That same day, he got a call from the Theranos lawyer David Doyle. After trying for weeks to get the Boies Schiller attorneys to propose a date for Ian's deposition, the Fuiszes' lawyers had run out of patience and sent notice that he would have to appear at their offices in Campbell, California, at 9:00 a.m. on May 17.

That's what Doyle was calling about. With the deadline for his appearance less than two days away, the lawyer encouraged Ian to invoke health issues to get out of the deposition and emailed him a doctor's note for his physician to adapt and sign. Ian forwarded the email to his personal Gmail address and, from there, to his wife's email address, asking her to print it. His anxiety seemed to reach a new fevered pitch.

Rochelle had known for a while that Ian wasn't well, but she'd had other concerns weighing on her mind: she was grieving her mother, who had just passed away and left behind a complicated estate to sort out, and she had just launched a new law practice with an associate. Part of her had been resentful that she wasn't getting the marital support she needed during this stressful period of her life. But

Ian's anguished state that day made her realize how dire his mental condition had become. She got him to agree to get help and scheduled an appointment with his general practitioner for the next morning.

WHEN ROCHELLE GOT UP around seven thirty a.m. on May 16, she saw that the bathroom light was on and the door closed. She assumed Ian was getting ready to go to the doctor's. But when he failed to come out after a while and didn't answer her calls, she pushed the bathroom door open. She found her husband hunched over in a chair unconscious and barely breathing. Panicked, she called 911.

Ian spent the next eight days hooked up to a ventilator at Stanford Hospital. He had taken enough acetaminophen, the active ingredient in painkillers like Tylenol, to kill a horse. Combined with the wine he'd consumed, the drug had destroyed his liver. He was pronounced dead on May 23. As an expert chemist, Ian knew exactly what he was doing. Rochelle later found a signed will that he'd had witnessed by Paul Patel and another colleague a few weeks before.

Rochelle was overwhelmed with grief but she found the strength to call Elizabeth's office and left a message with her assistant informing her about Ian's passing. Elizabeth didn't call back. Instead, later that day, Rochelle received an email from a

Theranos lawyer requesting that she immediately return Ian's company laptop and cell phone and any other confidential information he might have retained.

Inside Theranos, Ian's death was handled with the same cold, businesslike approach. Most employees weren't even informed of it. Elizabeth notified only a small group of company veterans in a brief email that made a vague mention of holding a memorial service for him. She never followed up and no service was held. Longtime colleagues of Ian's like Anjali Laghari, a chemist who had worked closely with him for eight years at Theranos and for two years before that at another biotech company, were left guessing about what had happened. Most thought he had died of cancer.

Tony Nugent became upset that nothing was done to honor his late colleague's memory. He and Ian hadn't been close. In fact, they had fought like cats and dogs at times during the Edison's development. But he was bothered by the lack of empathy being shown toward someone who had contributed nearly a decade of his life to the company. It was as if working at Theranos was gradually stripping them all of their humanity. Determined to show he was still a human being with compassion for his fellow man, Tony downloaded a list of Ian's patents from the patent office's online database and cut and pasted them into an email. He embedded a photo of Ian above the list and sent the email around to the two dozen colleagues he could think of who had worked with him, making a point to copy Elizabeth. It wasn't much, but it would at least give people something to remember him by, Tony thought.

| THIRTEEN |

Chiat\Day

"You're the leader." **Click. Click. Click.** "Strong, powerful." **Click. Click.** "Think of your mission." **Click. Click. Click. Click.**

The famous portrait photographer Martin Schoeller was softly whispering directions to Elizabeth in his thick German accent to elicit a range of emotions from her as he snapped her picture. She was wearing a thin black turtleneck and red lipstick, her hair brushed back in a loose bun that covered her ears. Two vertical lamps were set on either side of the chair she was sitting in to flatly illuminate her narrow face and create the white lights in her pupils that were a trademark of Schoeller's photographs.

Hiring Schoeller had been the idea of Patrick

O'Neill, the creative director of advertising agency TBWA\Chiat\Day's Los Angeles office. Chiat\Day was working on a secret marketing campaign for Theranos. The assignment ranged from creating a brand identity to building a new website and a smartphone app for the company ahead of the commercial launch of its blood-testing services in Walgreens and Safeway stores.

Elizabeth had chosen Chiat\Day because it was the agency that represented Apple for many years, creating its iconic 1984 Macintosh ad and later its "Think Different" campaign in the late 1990s. She'd even tried to convince Lee Clow, the creative genius behind those ads, to come out of retirement to work for her. Clow had politely referred her back to the agency, where she had immediately connected with Patrick.

A strikingly handsome man with blond hair, blue eyes, and the sculpted physique of someone who worked out religiously, Patrick was taken with Elizabeth from the moment he met her. His attraction wasn't of a romantic nature; he was gay. Rather, he was drawn in by her charisma and her singular drive to put a dent in the universe. He'd worked at Chiat\Day for fifteen years, creating ads for big corporate clients like Visa and IKEA. The work was interesting, but it didn't inspire him the way Elizabeth had when she'd first come to the agency's converted warehouse in Playa del Rey and described the Theranos mission of giving people access to pain-free, low-cost health care. In advertising, it wasn't often you got to work on something that really had the potential to make the world better. Patrick hadn't been surprised or put off by Theranos's insistence on absolute secrecy. Apple had been the same way. He understood technology companies' need to protect their valuable intellectual property. In any case, the company would soon be coming out of "stealth mode," as Elizabeth called it, and that's where he came in: his job was to make its commercial launch as impactful as possible.

Redesigning the Theranos website was a big part of that. Schoeller's photos were going to be featured on it. Not just those of Elizabeth. The photographer had spent most of the two-day shoot at a studio in Culver City taking pictures of models posing as patients. They were of different ages, genders, and ethnicities: children under five, children between five and ten, young men and women, middle-aged folks, and seniors. Some were white, some black, others Hispanic or Asian. The message was that Theranos's blood-testing technology would help everyone.

Elizabeth and Patrick spent hours choosing which patient photos to use from the shoot. Elizabeth wanted the faces displayed on the website to communicate empathy. She talked movingly about the sadness people felt when they found out that a loved one was sick and that it was too late to do anything about it. Theranos's painless blood tests would change that by catching diseases early, before they became a death sentence, she said.

PATRICK AND A GROUP of his Chiat\Day colleagues had been flying up to Palo Alto once a week to brainstorm with Elizabeth, Sunny, and her brother Christian throughout the fall of 2012 and into the winter and spring of 2013—the period when Ian Gibbons was spiraling into depression and Steve Burd was serving out his last months as Safeway CEO. Elizabeth scheduled the meetings on Wednesdays after learning that Apple's creative meetings with the agency had always been that day of the week. She told Patrick she admired the simplicity of Apple's brand message and wanted to emulate it.

Inside Chiat\Day, the Theranos assignment was known as "Project Stanford." Joining Patrick on the weekly trips up to Palo Alto were Carisa Bianchi, the president of the agency's L.A. office; Lorraine Ketch, the agency's chief of strategy; Stan Fiorito, who oversaw the account; and Mike Yagi, a copywriter. Early on, the Chiat\Day team decided that the best visual representation of Theranos's innovation was the miniature vial it had created to collect blood from fingertips. Elizabeth called it the "nanotainer." It was fitting because it really was tiny. At 1.29 centimeters, it was shorter than a dime positioned upright. Patrick wanted to take pictures of it to convey its scale to doctors and patients. But Elizabeth and Sunny were very concerned that word of it might leak out before the launch if anyone from the outside was allowed to see it. So they agreed that Chiat\Day would use its in-house photographer to take photos of it in the little photography studio the agency had in its Playa del Rey warehouse.

On the appointed day, Dan Edlin, one of Christian's Duke friends, flew down to L.A. with a custom-made plastic case containing twelve nanotainers. Checking it in with his bag at the airport was out of the question; it remained in his possession the whole flight down. When he got to the warehouse, Dan didn't let the little tubes out of his sight. No one at the agency was allowed to touch them except Patrick, who held one briefly and marveled at how small it was.

Real blood tended to turn purple after a while when it was exposed to air, so they filled one of the nanotainers with fake Halloween blood and took pictures of it against a white background. Patrick then made a photo montage showing it balancing on the tip of a finger. As he'd anticipated, it made for an arresting visual. Mike Yagi tried out different slogans to go with it, eventually settling on two that Elizabeth really liked: "One tiny drop changes everything" and "The lab test, reinvented." They blew the photo up and turned it into a mock full-page ad in the **Wall Street Journal**. In advertising lingo, this was known as a "tip-in." Elizabeth loved it and ordered a dozen more versions of it. She didn't say what she wanted them for, but Stan Fiorito got the sense she was using them as props during meetings with her board.

Patrick also worked with Elizabeth on a new company logo. Elizabeth believed in the Flower of Life, a geometric pattern of intersecting circles within a larger circle that pagans once considered the visual expression of the life that runs through all sentient beings. It was later adopted by the 1970s New Age movement as "sacred geometry" that provided enlightenment to those who spent time studying it.

The circle thus became the guiding motif of the Theranos brand. The inside of the "o" in "Theranos" was painted green to make it stand out, and the photos of the patient faces and of the nanotainer balancing on a fingertip were framed by circles. Patrick also created a new font for the website and marketing materials derived from Helvetica in which the dots over the "i" and the "j" and the periods at the end of sentences were round instead of square. Elizabeth seemed pleased with the results.

WHILE PATRICK REMAINED entranced with Elizabeth, Stan Fiorito was more circumspect. A gregarious ad industry veteran with reddish blond hair and freckles, Stan thought there was something odd about Sunny. He used a lot of software engineering jargon in their weekly meetings that had no applicability whatsoever to their marketing discussions. And when Stan tried to get him to walk him through how he'd arrived at what seemed like extremely aggressive sales targets, Sunny gave vague and boastful answers. Normally, companies did research to determine the size of the audience they were marketing to and then worked out what percentage of that audience they could realistically hope to convert into customers. But such basic concepts seemed lost on Sunny. Stan tried to look him up on the internet but couldn't find anything. He thought it was strange that someone with his background—a tech entrepreneur who had sold a company during the dot-com boom and made a lot of money in the process—had left no trace on the web. He wondered if Sunny had hired someone to scrub it for him.

It was also highly unusual for an obscure startup to hire a big ad agency like Chiat\Day. With their overhead and staffing, the big agencies were expensive. Chiat\Day was charging Theranos an annual retainer of $6 million a year. Where was this company nobody had heard of before getting the money to pay these types of fees? Elizabeth had stated on several occasions that the army was using her technology on the battlefield in Afghanistan and that it was saving soldiers' lives. Stan wondered if Theranos was funded by the Pentagon.

That would help explain the level of secrecy. Per

Sunny's instructions, any materials Theranos provided to Chiat\Day had to be numbered, logged, and kept in a locked room that only the team assigned to work on the account had access to. Any printing had to be done on a dedicated printer inside the room. Discarded materials couldn't just be thrown away, they had to be shredded. Computer files had to be stored on a separate server and could only be shared among the team via a dedicated intranet. And under no circumstances were they to share information about Theranos with anyone from Chiat\Day's L.A. office or the agency's broader network who hadn't signed a confidentiality agreement.

In addition to Mike Yagi, Stan supervised two other Chiat\Day employees who worked on the Theranos account full-time, Kate Wolff and Mike Peditto. Kate was in charge of building the website, while Mike was responsible for creating in-store brochures, signs, and an interactive iPad sales tool Theranos planned to use to pitch doctors.

As the months passed, Kate and Mike also began to develop concerns about their strange and demanding client. Both were from the East Coast and brought no-nonsense attitudes to their jobs. Kate, who was twenty-eight, grew up in Lincoln, Massachusetts, and played ice hockey at Boston University. Her proper, small-town upbringing had given her a strong moral compass. She also knew a thing or two about medicine: her dad and her wife were both doctors. Mike, who was thirty-two, was an Italian American from Philadelphia with a cynical streak who ran track and cross-country in college and graduate school. People didn't bullshit or take kindly to it where he came from.

Elizabeth wanted the website and all the various marketing materials to feature bold, affirmative statements. One was that Theranos could run "over 800 tests" on a drop of blood. Another was that its technology was more accurate than traditional lab testing. She also wanted to say that Theranos test results were ready in less than thirty minutes and that its tests were "approved by FDA" and "endorsed by key medical centers" such as the Mayo Clinic and the University of California, San Francisco's medical school, using the FDA, Mayo Clinic, and UCSF logos.

When she inquired about the basis for the claim about Theranos's superior accuracy, Kate learned that it was extrapolated from a study that had concluded that 93 percent of lab mistakes were due to human error. Theranos argued that, since its testing process was fully automated inside its device, that was grounds enough to say that it was more accurate than other labs. Kate thought that was a big leap in logic and said so. After all, there were laws against misleading advertising.

Mike felt the same way. In an email to Kate listing items that needed legal review, he included

"Automation makes us more accurate" and wrote in parentheses next to it, "this sounds like a puffery claim." Mike had never worked on a marketing campaign that involved medicine before and wanted to proceed extra carefully. Usually, healthcare campaigns, such as those involving pharmaceutical companies, were handled out of New York by a special division of the agency called TBWA\Health. He wondered why it wasn't handling this assignment or at least being consulted.

Elizabeth had mentioned a report several hundred pages long supporting Theranos's scientific claims. Kate and Mike repeatedly asked to see it, but Theranos wouldn't produce it. Instead, the company sent them a password-protected file containing what it said were excerpts from the report. It stated that the Johns Hopkins University School of Medicine had conducted due diligence on the Theranos technology and found it "novel and sound" and capable of "accurately" running "a wide range of routine and special assays."

Those quotes weren't from any lengthy report, however. They were from the two-page summary of Elizabeth and Sunny's meeting with five Hopkins officials in April 2010. As it had done with Walgreens, Theranos was again using that meeting to claim that its system had been independently evaluated. But that simply wasn't true. Bill Clarke, the director of clinical toxicology at the Johns Hopkins Hospital and one of three university scientists who attended the 2010 meeting, had asked Elizabeth to ship one of her devices to his lab so he could put it through its paces and compare its performance to the equipment he normally used. She had indicated she would but had never followed through. Kate and Mike didn't know any of this, but the fact that Theranos refused to show them the full report made them suspicious.

TO GAIN some insights into how to market to doctors, Chiat\Day proposed conducting focus group interviews with a few physicians. Theranos signed off on the idea but wanted to keep things hush-hush, so Kate enlisted her wife and her father to participate in them.

Kate's wife, Tracy, was chief resident at Los Angeles County General, where she was completing a residency in internal medicine and pediatrics. During her interview, which was conducted by phone, Tracy asked some questions that no one on the Theranos end of the line seemed able to answer. That evening, she told Kate she was dubious that the company had any truly novel technology. She questioned especially the notion that you could get enough blood from a finger to run tests accurately. Tracy's skepticism gave Kate pause.

Kate and Mike's main contacts at Theranos were Christian Holmes and his two Duke fraternity brothers, Dan Edlin and Jeff Blickman. Mike called them the "Therabros." He and Kate talked on the phone and exchanged emails with them frequently in the lead-up to the website launch. Theranos had initially wanted to go live with the site on April 1, 2013, but pushed the date back several times. A new launch date was set for September, but as the new deadline approached and Kate and Mike pressed the Therabros for substantiation to back up the claims Elizabeth wanted to make, it became apparent to them that some were exaggerated. For instance, they gleaned that Theranos couldn't produce tests results in less than thirty minutes. Kate watered that statement down to say that the results were ready in "4 hours or less," which she still had her doubts about. Kate and Mike also began to suspect that Theranos couldn't perform all its blood tests from small finger-stick draws and that it used traditional venous draws for some tests. They suggested adding a disclaimer to the website to make that clear, but the response they got from Christian and Jeff was that Elizabeth didn't want a disclaimer.

Mike was getting worried about Chiat\Day's legal liability. He went back and read the agency's contract with Theranos. It indemnified Chiat\Day for any claims made in marketing materials that the client had approved in writing. He fired off an email to Joe Sena, an attorney at the agency's outside law firm, Davis & Gilbert, asking if Chiat\Day should make Theranos use specific language in its written approvals. Sena replied that it wasn't necessary but urged him to keep track of those written approvals.

Kate, meanwhile, was sparring with Christian and Jeff about a line Elizabeth wanted to add to the website that read, "Send samples to us." Kate asked them what logistical system the company had in place to transport the blood samples from physicians' offices to its lab and eventually concluded it didn't have any. Doctors who "registered" for the service would merely be generating an automated email routed to Jeff's in-box. What would happen after that was anyone's guess. No one at Theranos had bothered to think it through, as far as Kate could tell.

THE FORTY-EIGHT HOURS before the website went live turned into a mad scramble. Mike Yagi, who for months had been laboring to write and rewrite copy to Elizabeth's satisfaction and was under a lot of stress, had an anxiety attack and went home to rest. He left the office so abruptly and in such a state that his colleagues didn't know whether he'd ever be coming back.

Then, on the evening before the launch, Theranos sent word that it wanted to have an emergency conference call. Kate, Mike, Patrick, Lorraine Ketch, and a copywriter named Kristina Altepeter who was filling in for Yagi gathered in the warehouse's board room (so named because its table was made from surfboards) and listened as Elizabeth announced that Theranos's legal team had ordered last-minute wording changes. Kate and Mike were annoyed. They'd been requesting a legal review for months. Why was it only happening now?

The call dragged on for more than three hours until 10:30 p.m. They went over the site line by line, as Elizabeth slowly dictated every alteration that needed to be made. Patrick nodded off at one point. But Kate and Mike stayed alert enough to notice that the language was being systematically dialed back. "Welcome to a revolution in lab testing" was changed to "Welcome to Theranos." "Faster results. Faster answers" became "Fast results. Fast answers." "A tiny drop is all it takes" was now "A few drops is all it takes."

A blurb of text next to the photo of a blond-haired, blue-eyed toddler under the headline "Goodbye, big bad needle" had previously referred only to finger-stick draws. Now it read, "Instead of a huge needle, we can use a tiny finger stick or collect a micro-sample from a venous draw." It wasn't lost on Kate and Mike that this was tantamount to the disclaimer they had previously suggested.

In a part of the site titled "Our Lab," a banner running across the page beneath an enlarged photo of a nanotainer had stated, "At Theranos, we can perform all of our lab tests on a sample 1/1,000 the size of a typical blood draw." In the new version of the banner, the words "all of" were gone. Lower down on the same page was the claim Kate had pushed back against months earlier. Under the heading "Unrivaled accuracy," it cited the statistic about 93 percent of lab errors being caused by humans and inferred from it that "no other laboratory is more accurate than Theranos." Sure enough, that was walked back too.

THE LAST-MINUTE REVISIONS only served to reinforce Kate and Mike's suspicions. Elizabeth had wanted all those sweeping claims to be true, but just because you badly wanted something to be real didn't make it so, Mike thought. He and Kate were beginning to question whether Theranos had any technology at all. Did its vaunted "black box," as people at Chiat\Day referred to the Theranos device, even exist?

They shared their mounting doubts with Stan, whose own interactions with Sunny were becoming increasingly unpleasant. Every quarter, Stan was having to chase Sunny around for money. Sunny kept asking him to justify the bills the agency submitted. Stan spent hours going over them with him point by point. Sunny would put him on speaker and pace around his office. When Stan asked him to move closer to the phone so he could make out what he was saying, Sunny's temper would flare.

Not everyone at Chiat\Day was souring on Theranos, however. The L.A. office's two higher-ups,

Carisa and Patrick, remained smitten with Elizabeth. Patrick idolized Lee Clow and the marketing magic he had conjured up for Apple. It was clear he thought Theranos had the potential to become his own big legacy moment. Kate voiced her concerns to him on several occasions, but he dismissed them as Kate just being Kate. She had a tendency to be overly dramatic, Patrick thought. His view was that she and Mike should stop questioning everything and just complete the work they were being asked to do. In Patrick's experience, all tech startups were chaotic and secretive. He saw nothing unusual or worrisome in that.

| FOURTEEN |

Going Live

Alan Beam was sitting in his office reviewing lab reports when Elizabeth poked her head in and asked him to follow her. She wanted to show him something. They stepped outside the lab into an area of open office space where other employees had gathered. At her signal, a technician pricked a volunteer's finger, then applied a transparent plastic implement shaped like a miniature rocket to the blood oozing from it. This was the Theranos sample collection device. Its tip collected the blood and transferred it to two little engines at the rocket's base. The engines weren't really engines: they were nanotainers. To complete the transfer, you pushed the nanotainers into the belly of the plastic rocket like a plunger. The movement created a vacuum that sucked the blood into them.

Or at least that was the idea. But in this instance, things didn't go quite as planned. When the technician pushed the tiny twin tubes into the device, there was a loud **pop** and blood splattered everywhere. One of the nanotainers had just exploded.

Elizabeth looked unfazed. "OK, let's try that again," she said calmly.

Alan wasn't sure what to make of the scene. He'd only been working at Theranos for a few weeks and was still trying to get his bearings. He knew the nanotainer was part of the company's proprietary blood-testing system, but he'd never seen one in action before. He hoped this was just a small mishap that didn't portend bigger problems.

The lanky pathologist's circuitous route to Silicon Valley had started in South Africa, where he grew up. After majoring in English at the University of the Witwatersrand in Johannesburg ("Wits" to South Africans), he'd moved to the United States to take premed classes at Columbia University in New York City. The choice was guided by his conservative Jewish parents, who considered only a few professions acceptable for their son: law, business, and medicine.

Alan had stayed in New York for medical school, enrolling at the Mount Sinai School of Medicine on Manhattan's Upper East Side, but he quickly realized that some aspects of being a doctor didn't suit his temperament. Put off by the crazy hours and the sights and smells of the hospital ward, he gravitated toward the more sedate specialty of laboratory science, which led to postdoctoral studies in virology and a residency in clinical pathology at Brigham and Women's Hospital in Boston.

In the summer of 2012, Alan was running the lab of a children's hospital in Pittsburgh when he noticed a job posting on LinkedIn that dovetailed perfectly with his budding fascination with Silicon Valley: laboratory director at a Palo Alto biotech firm. He had just finished reading Walter Isaacson's biography of Steve Jobs. The book, which he'd found hugely inspiring, had cemented his desire to move out to the San Francisco Bay Area.

After he applied for the job, Alan was asked to fly out for an interview scheduled for 6:00 p.m. on a Friday. The timing seemed odd but he was happy to oblige. He met with Sunny first and then with Elizabeth. There was something about Sunny that he found vaguely creepy, but that impression was more than offset by Elizabeth, who came off as very earnest in her determination to transform health care. Like most people who met her for the first time, Alan was taken aback by her deep voice. It was unlike anything he'd heard before.

Although he received a job offer just a few days later, Alan wasn't able to start working at Theranos immediately. First, he had to apply for his California medical license. That took eight months, delaying his official start until April 2013. By then, it had been almost a year since his predecessor, Arnold Gelb, had quit. In between, a semiretired lab director named Spencer Hiraki had come in occasionally to review and sign off on lab reports. That hadn't seemed too problematic to Alan since the Theranos lab was only testing a few samples a week from Safeway's employee clinic.

What did seem more troubling was the lab's morale when he took it over. Its members were down right despondent. During Alan's first week on the job, Sunny summarily fired one of the CLSs. The poor fellow was frog-marched out by security in front of everyone. Alan got the distinct impression it wasn't the first time something like that had happened. No wonder people's spirits were low, he thought.

The lab Alan inherited was divided into two parts: a room on the building's second floor that was filled with commercial diagnostic equipment, and a second room beneath it where research was being conducted. The upstairs room was the CLIA-certified part of the lab, the one Alan was responsible for. Sunny and Elizabeth viewed its conventional machines as dinosaurs that would soon be rendered extinct by Theranos's revolutionary technology, so they called it "Jurassic Park." They called the downstairs room "Normandy" in reference to the D-day landings during World War II. The proprietary Theranos devices it contained would take the lab industry by storm, like the Allied troops who braved hails of machine-gun fire on Normandy's beaches to liberate Europe from Nazi occupation.

In his eagerness and excitement, Alan initially bought into the bravado. But a conversation he had with Paul Patel shortly after the botched nanotainer demonstration raised questions in his mind about how far along the Theranos technology really was. Patel was the biochemist who led the development of blood tests for Theranos's new device, which Alan knew only by its code name—"4S." Patel let slip that his team was still developing its assays on lab plates on the bench. That surprised Alan, who had assumed the assays were already integrated into the 4S. When he asked why that wasn't the case, Patel replied that the new Theranos box wasn't working.

BY THE SUMMER of 2013, as Chiat\Day scrambled to ready the Theranos website for the company's commercial launch, the 4S, aka the miniLab, had been under development for more than two and a half years. But the device remained very much a work in progress. The list of its problems was lengthy.

The biggest problem of all was the dysfunctional corporate culture in which it was being developed. Elizabeth and Sunny regarded anyone who raised a concern or an objection as a cynic and a naysayer. Employees who persisted in doing so were usually marginalized or fired, while sycophants were promoted. Sunny had elevated a group of ingratiating Indians to key positions. One of them was Sam Anekal, the manager in charge of integrating the various components of the miniLab who had clashed with Ian Gibbons. Another was Chinmay Pangarkar, a bioengineer with a Ph.D. in chemical engineering from the University of California, Santa Barbara. There was also Suraj Saksena, a clinical chemist who had a Ph.D. in biochemistry and biophysics from Texas A&M. On paper, all three had impressive educational credentials, but they shared two traits: they had very little industry experience, having joined the company not long after finishing their studies, and they had a habit of telling Elizabeth and Sunny what they wanted to hear, either out of fear or out of desire to advance, or both.

For the dozens of Indians Theranos employed, the fear of being fired was more than just the dread of losing a paycheck. Most were on H-1B visas and dependent on their continued employment at the company to remain in the country. With a despotic boss like Sunny holding their fates in his hands, it was akin to indentured servitude. Sunny, in fact, had the master-servant mentality common among an older generation of Indian businessmen. Employees were his minions. He expected them to be at his disposal at all hours of the day or night and on weekends. He checked the security logs every morning to see when they badged in and out.

Every evening, around seven thirty, he made a fly-by of the engineering department to make sure people were still at their desks working.

With time, some employees grew less afraid of him and devised ways to manage him, as it dawned on them that they were dealing with an erratic man-child of limited intellect and an even more limited attention span. Arnav Khannah, a young mechanical engineer who worked on the miniLab, figured out a surefire way to get Sunny off his back: answer his emails with a reply longer than five hundred words. That usually bought him several weeks of peace because Sunny simply didn't have the patience to read long emails. Another strategy was to convene a biweekly meeting of his team and invite Sunny to attend. He might come to the first few, but he would eventually lose interest or forget to show up.

While Elizabeth was fast to catch on to engineering concepts, Sunny was often out of his depth during engineering discussions. To hide it, he had a habit of repeating technical terms he heard others using. During a meeting with Arnav's team, he latched onto the term "end effector," which signifies the claws at the end of a robotic arm. Except Sunny didn't hear "end effector," he heard "endofactor." For the rest of the meeting, he kept referring to the fictional endofactors. At their next meeting with Sunny two weeks later, Arnav's team brought a PowerPoint presentation titled "Endofactors Update." As Arnav flashed it on a screen with a projector, the five members of his team stole furtive glances at one another, nervous that Sunny might become wise to the prank. But he didn't bat an eye and the meeting proceeded without incident. After he left the room, they burst out laughing.

Arnav and his team also got Sunny to use the obscure engineering term "crazing." It normally refers to a phenomenon that produces fine cracks on the surface of a material, but Arnav and his colleagues used it liberally and out of context to see if they could get Sunny to repeat it, which he did. Sunny's knowledge of chemistry was no better. He thought the chemical symbol for potassium was P (it's K; P is the symbol for phosphorus)—a mistake most high school chemistry students wouldn't make.

Not all the setbacks encountered during the miniLab's development could be laid at Sunny's feet, however. Some were a consequence of Elizabeth's unreasonable demands. For instance, she insisted that the miniLab cartridges remain a certain size but kept wanting to add more assays to them. Arnav didn't see why the cartridges couldn't grow by half an inch since consumers wouldn't see them. After her run-in with Lieutenant Colonel David Shoemaker, Elizabeth had abandoned her plan of putting the Theranos devices in Walgreens stores and operating them remotely, to avoid problems with the FDA. Instead, blood pricked from patients' fingers would be couriered to Theranos's Palo Alto lab and tested there. But she remained stuck on the notion that the miniLab was a consumer device, like an iPhone or an iPad, and that its components needed to look small and pretty. She still nurtured the ambition of putting it in people's homes someday, as she had promised early investors.

Another difficulty stemmed from Elizabeth's insistence that the miniLab be capable of performing the four major classes of blood tests: immunoassays, general chemistry assays, hematology assays, and assays that relied on the amplification of DNA. The only known approach that would permit combining all of them in one desktop machine was to use robots wielding pipettes. But this approach had an inherent flaw: over time, a pipette's accuracy drifts. When the pipette is brand new, aspirating 5 microliters of blood might require the little motor that activates the pipette's pump to rotate by a certain amount. But three months later, that exact same rotation of the motor might yield only 4.4 microliters of blood—a large enough difference to throw off the entire assay. While pipette drift was something that ailed all blood analyzers that relied on pipetting systems, the phenomenon was particularly pronounced on the miniLab. Its pipettes had to be recalibrated every two to three months, and the recalibration process put the device out of commission for five days.

Kyle Logan, a young chemical engineer who'd joined Theranos right out of Stanford after earning an academic award there named after Channing Robertson, had frequent debates with Sam Anekal about this issue. He thought the company should migrate to a more reliable system that didn't involve pipettes, such as the one Abaxis used in its Piccolo Xpress analyzer. Sam would reply that the Piccolo could perform only one class of blood test, general chemistry assays. (Unlike immunoassays, which measure a substance in the blood by using antibodies that bind to the substance, general chemistry assays rely on other chemical principles such as light absorbance or electrical signal changes.) Elizabeth wanted a machine that was more versatile, he'd remind Kyle.

Compared to big commercial blood analyzers, another one of the miniLab's glaring weaknesses was that it could process only one blood sample at a time. Commercial machines were bulky for a reason: they were designed to process hundreds of samples simultaneously. In industry jargon, this was known as having a "high throughput." If the Theranos wellness centers attracted a lot of patients, the miniLab's low throughput would result in long wait times and make a joke of the company's promise of fast test results.

In an attempt to remedy this problem, someone had come up with the idea of stacking six miniLabs one on top of the other and having them share one cytometer to reduce the size and cost of the resulting contraption. This Frankenstein machine was called the "six-blade," a term borrowed from the computer industry, where stacking servers on top of one another is common to save space and energy. In these modular stacking configurations, each server is referred to as a "blade."

But no one had stopped to consider what implications this design would have with respect to one key variable: temperature. Each miniLab blade generated heat, and heat rises. When the six blades were processing samples at the same time, the temperature in the top blades reached a level that interfered with their assays. Kyle, who was twenty-two and just out of college, couldn't believe something that basic had been overlooked.

Aside from its cartridge, pipette, and temperature issues, many of the other technical snafus that plagued the miniLab could be chalked up to the fact that it remained at a very early prototype stage. Less than three years was not a lot of time to design and perfect a complex medical device. These problems ranged from the robots' arms landing in the wrong places, causing pipettes to break, to the spectrophotometers being badly misaligned. At one point, the blood-spinning centrifuge in one of the miniLabs blew up. These were all things that could be fixed, but it would take time. The company was still several years away from having a viable product that could be used on patients.

However, as Elizabeth saw it, she didn't have several years. Twelve months earlier, on June 5, 2012, she'd signed a new contract with Walgreens that committed Theranos to launching its blood-testing services in some of the pharmacy chain's stores by February 1, 2013, in exchange for a $100 million "innovation fee" and an additional $40 million loan.

Theranos had missed that deadline—another postponement in what from Walgreens's perspective had been three years of delays. With Steve Burd's retirement, the Safeway partnership was already falling apart, and if she waited much longer, Elizabeth risked losing Walgreens too. She was determined to launch in Walgreens stores by September, come hell or high water.

Since the miniLab was in no state to be deployed, Elizabeth and Sunny decided to dust off the Edison and launch with the older device. That, in turn, led to another fateful decision—the decision to cheat.

IN JUNE, Daniel Young, the brainy MIT Ph.D. who headed Theranos's biomath team, came to see Alan Beam in Jurassic Park with a subordinate named Xinwei Gong in tow. In the five years since he'd joined Theranos, Daniel had risen up the ranks to become the company's de facto number-three executive. He had Elizabeth and Sunny's ear, and they often deferred to him to solve nettlesome technical problems.

In his first few years at Theranos, Daniel had

seemed every bit the family man, leaving the office at six every evening to have dinner with his wife and kids. This routine had drawn snickers behind his back from some colleagues. But after being promoted to vice president, Daniel had become a different person. He worked longer hours and stayed at the office later. He got very drunk at company parties, which was jarring because he was always quiet and inscrutable at work. And there were whispers of a flirtation with a coworker.

Daniel told Alan that he and Gong, who went by Sam, were going to tinker with the ADVIA 1800, one of the lab's commercial analyzers. The ADVIA was a hulking 1,320-pound machine the size of two large office copiers put together that was made by Siemens Healthcare, the German conglomerate's medical-products subsidiary.

Over the next few weeks, Alan observed Sam spend hours opening the machine up and filming its innards with his iPhone camera. He was hacking into it to try to make it compatible with small finger-stick blood samples, Alan realized. It seemed like confirmation of what Paul Patel had told him: the 4S must not be working, otherwise why resort to such desperate measures? Alan knew the Edison could only perform immunoassays, so it made sense that Daniel and Sam would choose the ADVIA, which specialized in general chemistry assays.

One of the panels of blood tests most commonly ordered by physicians was known as the "chem 18" panel. Its components, which ranged from tests to measure electrolytes such as sodium, potassium, and chloride to tests used to monitor patients' kidney and liver function, were all general chemistry assays. Launching in Walgreens stores with a menu of blood tests that didn't include these tests would have been pointless. They accounted for about two-thirds of doctors' orders.

But the ADVIA was designed to handle a larger quantity of blood than you could obtain by pricking a finger. So Daniel and Sam thought up a series of steps to adapt the Siemens analyzer to smaller samples. Chief among these was the use of a big robotic liquid handler called the Tecan to dilute the little blood samples collected in the nanotainers with a saline solution. Another was to transfer the diluted blood into custom-designed cups half the size of the ones that normally went into the ADVIA.

The combination of these two steps solved a problem known as "dead volume." Like many commercial analyzers, the ADVIA featured a probe that dropped down into the blood sample and aspirated it. Although it aspirated most of the sample, there was always some unused liquid left at the bottom. Reducing the sample cup's size brought its bottom closer to the probe's tip and diluting the blood created more liquid to work with.

Alan had reservations about the dilution part.

The Siemens analyzer already diluted blood samples when it performed its assays. The protocol Daniel and Sam had come up with meant that the blood would be diluted twice, once before it went into the machine and a second time inside it. Any lab director worth his salt knew that the more you tampered with a blood sample, the more room you introduced for error.

Moreover, this double dilution lowered the concentration of the analytes in the blood samples to levels that were below the ADVIA's FDA-sanctioned analytic measurement range. In other words, it meant using the machine in a way that neither the manufacturer nor its regulator approved of. To get the final patient result, one had to multiply the diluted result by the same factor the blood had been diluted by, not knowing whether the diluted result was even reliable. Daniel and Sam were nonetheless proud of what they'd accomplished. At heart, both were engineers for whom patient care was an abstract concept. If their tinkering turned out to have adverse consequences, they weren't the ones who would be held personally responsible. It was Alan's name, not theirs, that was on the CLIA certificate.

When their work was finished, a Theranos lawyer named Jim Fox came by Alan's office and suggested the company patent what they had done. That seemed to Alan like a ridiculous idea. To his mind, fiddling with another manufacturer's device didn't amount to inventing something new, especially if it no longer worked as well afterward.

Word that the Siemens machines had been jailbroken made its way to Ted Pasco, who had succeeded John Fanzio as procurement manager and in the process inherited his role as the prime recipient of company gossip. Ted soon got confirmation of what he'd heard through the rumor mill when he received instructions from Elizabeth and Sunny to purchase six more ADVIAs. He negotiated a bulk discount with Siemens, but the order still cost well north of $100,000.

As September 9, 2013, approached, the date Elizabeth had set for the launch, Alan grew worried that Theranos wasn't ready. Two of the assays performed on the hacked Siemens analyzers were giving the lab particular trouble: sodium and potassium. Alan suspected the cause of the latter was a phenomenon known as "hemolysis," which occurs when red blood cells burst and release extra potassium into the sample. Hemolysis was a known side effect of finger-stick collection. Milking blood from a finger put stress on red blood cells and could cause them to break apart.

Alan had noticed a piece of paper with a number on it taped to Elizabeth's office window. It was her launch countdown. The sight of it made him panic. A few days before the launch, he went to see her and asked her to delay. Elizabeth wasn't her usual confident self. Her voice was tremulous and she was visibly shaking as she tried to reassure him that everything would be OK. If necessary, they could fall back on regular venous draws, she told him. That briefly made Alan feel better, but his anxiety returned as soon as he left her office.

ANJALI LAGHARI, the chemist who had worked with Ian Gibbons for a total of ten years at Theranos and another biotech company, was dismayed when she returned from her three-week vacation in India in late August.

Anjali headed the immunoassay group. Her team had been trying for years to develop blood tests on Theranos's older device, the Edison. Much to her frustration, the black-and-white machines' error rate was still high for some tests. Elizabeth and Sunny had been promising her for a year that all would be well once the company introduced its next-generation device, the 4S. Except that day never seemed to arrive. That was fine as long as Theranos remained a research-and-development operation, which was still the case when Anjali had departed for India three weeks earlier. But now everyone was suddenly talking about "going live" and there were emails in her in-box referring to an imminent commercial launch.

Launch? With what? Anjali wondered with growing alarm.

In her absence, she learned, employees who were

not authorized CLIA lab personnel had been let into the lab. She didn't know why, but she did know the lab was under instructions to conceal whatever it was they were doing from Siemens representatives when they came by to service the German manufacturer's machines.

Changes had also been made to the way samples were being processed on the Edisons. Under Sunny's orders, they were now being prediluted with a Tecan liquid handler before being run through the device. This was to make up for the fact that the Edison could run at most three tests on one fingerstick sample. Prediluting the blood created more volume to run more tests. But if the device already had a high error rate under normal circumstances, an additional dilution step seemed likely only to make things worse. Anjali also had concerns about the nanotainers. Blood would dry up in the little tubes and she and her colleagues often couldn't extract enough from them.

She tried to talk sense into Elizabeth and Daniel Young by emailing them Edison data from Theranos's last study with a pharmaceutical company—Celgene—which dated back to 2010. In that study, Theranos had used the Edison to track inflammatory markers in the blood of patients who had asthma. The data had shown an unacceptably high error rate, causing Celgene to end the companies' collaboration. Nothing had changed since that failed study, Anjali reminded them.

Neither Elizabeth nor Daniel acknowledged her email. After eight years at the company, Anjali felt she was at an ethical crossroads. To still be working out the kinks in the product was one thing when you were in R&D mode and testing blood volunteered by employees and their family members, but going live in Walgreens stores meant exposing the general population to what was essentially a big unauthorized research experiment. That was something she couldn't live with. She decided to resign.

When Elizabeth heard the news, she asked Anjali to come by her office. She wanted to know why she was leaving and whether she could be persuaded to stay. Anjali repeated her concerns: the Edison's error rate was too high and the nanotainer still had problems. Why not wait until the 4S was ready? Why rush to launch now? she asked.

"Because when I promise something to a customer, I deliver," Elizabeth replied.

That response made no sense to Anjali. Walgreens was just a business partner. Theranos's ultimate customers would be the patients who came to Walgreens stores and ordered its blood tests thinking they could rely on them to make medical decisions. Those were the customers Elizabeth should be worrying about. By the time Anjali got back to her desk, word of her resignation had spread and coworkers were coming up to her to say goodbye. She had given one week's notice and was planning to work until the end of her notice period, but Sunny didn't like the scene these public farewells were causing. He sent Mona Ramamurthy, the head of human resources, to tell her that she had to leave immediately.

On her way out, Anjali printed the email she had sent to Elizabeth and Daniel. She had a feeling this wasn't going to end well and she needed something to protect herself, something that would show she had disagreed with the decision to launch. Forwarding the email to her personal Yahoo account would have been easier, but she knew Sunny closely monitored employees' email activity. So she hid the printout in her bag and smuggled it out. Anjali wasn't the only one with misgivings. Tina Noyes, her deputy in the immunoassay group who had worked at Theranos for more than seven years, also quit.

The resignations infuriated Elizabeth and Sunny. The following day, they summoned the staff for an all-hands meeting in the cafeteria. Copies of **The Alchemist,** Paulo Coelho's famous novel about an Andalusian shepherd boy who finds his destiny by going on a journey to Egypt, had been placed on every chair. Still visibly angry, Elizabeth told the gathered employees that she was building a religion. If there were any among them who didn't believe, they should leave. Sunny put it more bluntly: anyone not prepared to show complete devotion and unmitigated loyalty to the company should "get the fuck out."

FIFTEEN

Unicorn

She hated the artist's illustration. He had made her head huge and given her a vacuous, doe-eyed grin that screamed blond bimbo. Other than that, there wasn't much to dislike about the article. It took up most of a page in the front section of the **Wall Street Journal** and hit all the right notes. Drawing blood the traditional way with a needle in the arm was likened to vampirism, or as the writer put it more elegantly, "medicine by Bram Stoker." Theranos's processes, by contrast, were described as requiring "only microscopic blood volumes" and as "faster, cheaper and more accurate than the conventional methods." The brilliant young Stanford dropout behind the breakthrough invention was anointed "the next Steve Jobs or Bill Gates" by no less than former secretary of state George Shultz, the man many credited with winning the Cold War, in a quote at the end of the article.

Elizabeth had engineered the piece, which was published in the Saturday, September 7, 2013, edition of the **Journal,** to coincide with the commercial launch of Theranos's blood-testing services. A press release was due to go out first thing Monday morning announcing the opening of the first Theranos wellness center in a Walgreens store in Palo Alto and plans for a subsequent nationwide expansion of the partnership. For a heretofore unknown startup, coverage this flattering in one of the country's most prominent and respected publications was a major coup. What had made it possible was Elizabeth's close relationship with Shultz—a connection she'd made two years earlier and carefully cultivated.

The former statesman, who in addition to crafting the Reagan administration's foreign policy also served as secretary of labor and secretary of the treasury under President Nixon, had joined the Theranos board of directors in July 2011 and become one of Elizabeth's biggest champions. A distinguished fellow at the Hoover Institution, the think tank housed on the Stanford campus, Shultz remained a revered and influential figure in Republican circles despite his advancing age (he was ninety-two). That made him a friend of the **Journal**'s famously conservative editorial page, to which he occasionally contributed op-eds.

During a visit to the paper's Midtown Manhattan headquarters to discuss climate change with its editorial board in 2012, Shultz had dropped mention of a secretive and reclusive Silicon Valley startup founder who he felt certain was going to revolutionize medicine with her technology. Intrigued, the **Journal**'s long-serving editorial page editor, Paul Gigot, had offered to send one of his writers to interview the mysterious wunderkind when she felt ready to break her silence and introduce her invention to the world. A year later, Shultz had called back with word that Elizabeth was ready and Gigot had handed the assignment to Joseph Rago, a member of the **Journal**'s editorial board who had written extensively about health care. The resulting piece ran in the Weekend Interview column, a fixture of the Saturday **Journal**'s opinion pages.

Elizabeth had picked a safe place for her coming-out party. The Weekend Interview, which rotated among the members of Gigot's staff, wasn't meant to be hard-hitting investigative journalism. Rather, it was what its name implied: an interview whose tone was usually friendly and nonconfrontational. Moreover, her message of bringing disruption to an old and inefficient industry was bound to play well with the **Journal** editorial page's pro-business, anti-regulation ethos. Nor did Rago, who had won a Pulitzer Prize for tough editorials dissecting Obamacare, have any reason to suspect that what

Elizabeth was telling him wasn't true. During his visit to Palo Alto, she had shown him the miniLab and the six-blade side by side and he had volunteered for a demonstration, receiving what appeared to be accurate lab results in his email in-box before he even left the building. What he didn't know was that Elizabeth was planning to use the Walgreens launch and his accompanying article containing her misleading claims as the public validation she needed to kick-start a new fund-raising campaign, one that would propel Theranos to the forefront of the Silicon Valley stage.

MIKE BARSANTI WAS vacationing in Lake Tahoe when he received a call on his cell phone from Donald A. Lucas, the son of legendary venture capitalist Donald L. Lucas. Mike and Don had gone to college together in the early 1980s at Santa Clara University and had remained friendly ever since. Mike was the retired chief financial officer of a Bay Area seafood and poultry business that his family had operated for more than six decades before selling it the previous year.

Don was calling to pitch Mike on an investment: Theranos. That came as a surprise to Mike. He had last heard of the startup seven years earlier when he and Don had attended a twenty-minute presentation Elizabeth gave on Sand Hill Road showcasing her little blood-testing machine. Mike remembered

Elizabeth very well: she'd come off as a dowdy young scientist back then, wearing Coke-bottle glasses and no makeup, speaking nervously to an audience of men two to three times her age. At the time, Don ran RWI Ventures, a firm he'd founded in the mid-1990s after spending ten years learning the ropes of the venture capital business working for his father. Mike had been an investor in RWI. His curiosity piqued by the awkward but obviously smart young woman, he'd asked Don why the firm wasn't taking a flyer on her like his father had. Don had replied that after careful consideration he'd decided against it. Elizabeth was all over the place, she wasn't focused, his father couldn't control her even though he chaired her board, and Don didn't like or trust her, Mike recalled his friend telling him.

"Don, what's changed?" Mike now asked.

Don explained excitedly that Theranos had come a long way since then. The company was about to announce the launch of its innovative finger-stick tests in one of the country's largest retail chains. And that wasn't all, he said. The Theranos devices were also being used by the U.S. military.

"Did you know they're in the back of Humvees in Iraq?" he asked Mike.

Mike wasn't sure he'd heard right. "What?" he blurted out.

"Yeah, I saw them stacked up at Theranos's headquarters after they came back."

If all this was true, these were impressive developments, Mike thought.

Don had launched a new firm in 2009 called the Lucas Venture Group. In recognition of her longstanding relationship with his aging father, who was addled by the onset of Alzheimer's disease, Elizabeth was giving him the chance to invest in the company at a discount to the price other investors were going to be offered in a big forthcoming fund-raising round. Intent on seizing what he saw as a great opportunity, the Lucas Venture Group was raising money for two new funds, Don told Mike. One of them was a traditional venture fund that would invest in several companies, including Theranos. The second would be exclusively devoted to Theranos. Did Mike want in? If so, time was short. The transaction had to close by the end of September.

A few weeks later, on the afternoon of September 9, 2013, Mike received an email from Don with the subject line "Theranos-time sensitive" that contained more details. The email, which went out to people who like Mike had previously invested in Don's funds, provided links to the **Wall Street Journal** article and to the Theranos press release from that morning. The Lucas Venture Group, it said, had been "invited" to invest up to $15 million in Theranos. The discounted price Elizabeth was offering the firm valued the company at $6 billion.

Mike took a deep breath. That was a huge valuation. He couldn't help but be annoyed with Don. When his friend had brushed aside his suggestion that they invest seven years before, Theranos had been valued around $40 million, he remembered ruefully.

Granted, the company seemed like a much surer bet now. Don's email said Theranos had "signed contracts and partnerships with very large retailers and drug stores as well as various pharmaceutical companies, HMO's, insurance agencies, hospitals, clinics and various government agencies." It also said the company had been "cash flow positive since 2006."

Mike and ten of his family members had pooled their money together in a limited liability company so they could invest in these types of venture deals. After conferring with them, he decided to pull the trigger on the investment and sent Don $790,000. Dozens of other Lucas Venture Group investors, known in industry parlance as "limited partners," did the same, cutting checks of varying amounts. They ranged from Robert Colman, a co-founder of the defunct San Francisco investment bank Robertson Stephens & Co., to a retired Palo Alto psychotherapist.

BY THE FALL of 2013, money was flowing into the Valley ecosystem at such a dizzying pace that a new term was coined to describe the new breed of startups it was spawning. In an article published on the technology news website TechCrunch on November 2, 2013, a venture capitalist named Aileen Lee wrote about the proliferation of startups valued at $1 billion or more. She called them "unicorns." Despite their moniker, these tech unicorns were no myth: by Lee's count, there were thirty-nine of them—a number that would soon soar past one hundred.

Instead of rushing to the stock market like their dot-com predecessors had in the late 1990s, the unicorns were able to raise staggering amounts of money privately and thus avoid the close scrutiny that came with going public.

The poster child of the unicorns was Uber, the ride-hailing smartphone app cofounded by the hard-charging engineer Travis Kalanick. A few weeks before Elizabeth's **Journal** interview, Uber had raised $361 million at a valuation of $3.5 billion. There was also Spotify, the music streaming service that in November 2013 raised $250 million at a price per share that valued the whole company at $4 billion.

These companies' valuations would keep rising over the next few years, but for now they had been leapfrogged by Theranos. And the gap was about to get bigger.

The **Journal** article caught the attention of Christopher James and Brian Grossman, two seasoned investment professionals who ran a hedge fund based in San Francisco called Partner Fund Management. With about $4 billion in assets, Partner Fund had a successful track record, having notched up average annual returns of nearly 10 percent since James had founded it in 2004. Part of that success could be credited to its large health-care portfolio, which was overseen by Grossman.

After they reached out to her, Elizabeth invited James and Grossman over for a meeting on December 15, 2013. When they arrived at Theranos's headquarters, a sprawling beige structure built into the side of a hill a stone's throw from the Stanford campus, the first thing they both noticed was the heavy security. There were multiple guards at the entrance and they had to sign nondisclosure agreements just to be allowed into the building. Once inside, guards escorted them everywhere, even to the bathroom. Parts of the building couldn't be accessed without special key cards and were off-limits to them.

Elizabeth and Sunny had always been security conscious, but their level of paranoia had reached a new peak with the launch in Walgreens stores. They had convinced themselves that Quest and LabCorp viewed Theranos as a mortal threat to their cozy oligopoly and that they would try to quash their new competitor by any means available. There was also the matter of the promise John Fuisz had made in his deposition to "fuck with" Elizabeth until she died. She took that threat very seriously. Following his retirement from the military earlier that year, James Mattis had joined the Theranos board and, on his recommendation, Elizabeth had hired Jim Rivera, the head of his Pentagon security detail. Rivera was a grizzled pro who had protected Mattis during his frequent trips to Iraq and Afghanistan. He wore a gun holstered under his jacket or around his ankle at all times and led a team of a half dozen guards who were dressed in black suits and wore earpieces.

The tight security measures made an impression on James and Grossman. It called to mind the lengths to which the Coca-Cola Company went to guard its secret Coke formula and suggested to them that Theranos had very valuable intellectual property to protect. The representations Elizabeth and Sunny made to them cemented that belief.

During that first meeting, Elizabeth and Sunny told their guests that Theranos's proprietary finger-stick technology could perform blood tests covering 1,000 of the 1,300 codes laboratories used to bill Medicare and private health insurers, according to a lawsuit Partner Fund later filed against the company. (Many blood tests involve several billing codes, so the actual number of tests represented by those thousand codes was in the low hundreds.)

At a second meeting three weeks later, they showed them a PowerPoint presentation containing scatter plots purporting to compare test data from Theranos's proprietary analyzers to data from conventional lab machines. Each plot showed data points tightly clustered around a straight line that rose up diagonally from the horizontal x-axis. This indicated that Theranos's test results were almost perfectly correlated with those of the conventional machines. In other words, its technology was as accurate as traditional testing. The rub was that much of the data in the charts wasn't from the miniLab or even from the Edison. It was from other commercial blood analyzers Theranos had purchased, including one manufactured by a company located an hour north of Palo Alto called Bio-Rad.

Sunny also told James and Grossman that Theranos had developed about three hundred different blood tests, ranging from commonly ordered tests to measure glucose, electrolytes, and kidney function to more esoteric cancer-detection tests. He boasted that Theranos could perform 98 percent of them on tiny blood samples pricked from a finger and that, within six months, it would be able to do all of them that way. These three hundred tests represented 99 to 99.9 percent of all laboratory requests, and Theranos had submitted every single one of them to the FDA for approval, he said.

Sunny and Elizabeth's boldest claim was that the Theranos system was capable of running seventy different blood tests simultaneously on a single finger-stick sample and that it would soon be able to run even more. The ability to perform so many tests on just a drop or two of blood was something of a Holy Grail in the field of microfluidics. Thousands of researchers around the world in universities and industry had been pursuing this goal for more than two decades, ever since the Swiss scientist Andreas Manz had shown that the microfabrication techniques developed by the computer chip industry could be repurposed to make small channels that moved tiny volumes of fluids.

But it had remained beyond reach for a few basic reasons. The main one was that different classes of blood tests required vastly different methods. Once you'd used your micro blood sample to perform an immunoassay, there usually wasn't enough blood left for the completely different set of lab techniques a general chemistry or hematology assay required. Another was that, while microfluidic chips could handle very small volumes, no one had yet figured out how to avoid losing some of the sample during its transfer to the chip. Losing a little bit of the blood sample didn't matter much when it was large, but it became a big problem when it was tiny. To hear Elizabeth and Sunny tell it, Theranos had solved these and other difficulties—challenges that had bedeviled an entire branch of bioengineering research.

Besides Theranos's supposed scientific accomplishments, what helped win James and Grossman over was its board of directors. In addition to Shultz and Mattis, it now included former secretary of state Henry Kissinger, former secretary of defense William Perry, former Senate Arms Services Committee chairman Sam Nunn, and former navy admiral Gary Roughead. These were men with sterling, larger-than-life reputations who gave Theranos a stamp of legitimacy. The common denominator between all of them was that, like Shultz, they were fellows at the Hoover Institution. After befriending Shultz, Elizabeth had methodically cultivated each one of them and offered them board seats in exchange for grants of stock.

The presence of these former cabinet members, congressmen, and military officials on the board also lent credence to Elizabeth and Sunny's assertions that Theranos's devices were being used in the field by the U.S. military. James and Grossman thought that Theranos's finger-stick offerings in Walgreens and Safeway stores were likely to be a hit with consumers and to capture a large share of the U.S. blood-testing market. A contract with the Department of Defense would add another big source of revenues.

A spreadsheet with financial projections Sunny sent the hedge fund executives supported this notion. It forecast gross profits of $165 million on revenues of $261 million in 2014 and gross profits of $1.08 billion on revenues of $1.68 billion in 2015. Little did they know that Sunny had fabricated these numbers from whole cloth. Theranos hadn't had a real chief financial officer since Elizabeth had fired Henry Mosley in 2006. The closest thing the company had to one was a corporate controller named Danise Yam. Six weeks after Sunny sent Partner Fund his projections, Yam sent very different ones to an advisory firm called Aranca for the purpose of pricing stock options for employees. Yam forecast a profit of $35 million in 2014 and of $100 million in 2015 ($130 million and $980 million less, respectively, than what Sunny projected to Partner Fund). The gap in revenues was even bigger: she predicted revenues of $50 million in 2014 and of $134 million in 2015 ($211 million and $1.55 billion less than the projections given to Partner Fund). As it would turn out, even Yam's numbers were wildly optimistic.

James and Grossman of course had no way of knowing that Theranos's internal projections were five- to twelvefold lower than the ones they were shown. It didn't occur to them that anything untoward could be going on at a company with such a prestigious board. Not to mention the fact that this board had a special adviser named David Boies who attended all of its meetings. With one of the country's top lawyers keeping watch, what could possibly go wrong?

On February 4, 2014, Partner Fund purchased 5,655,294 Theranos shares at a price of $17 a share—$2 a share more than the Lucas Venture Group had paid just four months earlier. The investment brought in another $96 million to Theranos's coffers and valued it at a stunning $9 billion. This meant that Elizabeth, who owned slightly more than half of the company, now had a net worth of almost $5 billion.

| SIXTEEN |

The Grandson

Standing in the middle of a crowd of his new colleagues in the cafeteria of the old Facebook building, Tyler Shultz listened to an emotional speech Elizabeth was giving. She was talking about her uncle's premature death from cancer and how an early warning from Theranos's blood tests could have prevented it. That was what she had spent the past ten years tirelessly working toward, she said teary-eyed, her voice catching: a world in which no one would have to say goodbye to a loved one too soon. Tyler found the message deeply inspiring. He had started working at Theranos less than a week earlier, after graduating from Stanford the previous spring and taking the summer off to backpack around Europe. There had been a lot to absorb in the space of a few days, not least of which was the news Elizabeth had called this all-employee meeting to announce: the company was going live with its technology in Walgreens stores.

Tyler had first met Elizabeth in late 2011 when he'd dropped by his grandfather George's house near the Stanford campus. He was a junior then, majoring in mechanical engineering. Elizabeth's vision of instant and painless tests run on drops of blood collected from fingertips had struck an immediate chord with him. After interning at Theranos that summer, he'd changed his major to biology and applied for a full-time position at the company.

His first day at work had been filled with drama. A woman named Anjali who headed the immunoassay team had quit, and a group of employees had gathered in the parking lot to say goodbye to her. Word was that Anjali and Elizabeth had had a big falling-out. Then, three days later, Tyler had been informed that the protein engineering group he'd originally been assigned to was being disbanded and everyone was being moved to the undermanned immunoassay team to help out. It was all a bit chaotic and confusing, but Elizabeth's stirring speech made his budding concerns melt away. He left the meeting energized and motivated to work really hard.

A month into the job, Tyler met a new hire named Erika Cheung. Like Tyler, Erika was a newly minted college grad who'd majored in biology, but that was about all they had in common. With his dirty blond hair and famous grandfather, Tyler was a product of the establishment, while Erika came from a middle-class mixed-race family. Her father had emigrated to the United States from Hong Kong and worked his way up from package handler to engineering manager at UPS. She'd spent large stretches of her youth homeschooled.

Despite their very different backgrounds, Tyler and Erika became fast friends. Their job on the immunoassay team was to help run experiments to verify the accuracy of blood tests on Theranos's Edison devices before they were deployed in the lab for use on patients. This verification process was known as "assay validation." The blood samples used for these experiments came from employees and sometimes from employees' friends and family members. To encourage employees to give blood, Theranos paid them $10 per tube. That meant you could make as much as $50 in one sitting. Tyler and Erika competed to see who could get to $600 first—the threshold beyond which the company had to report the payments as compensation to the IRS. One weekend, Theranos was looking for more volunteers, so Tyler recruited his four housemates to come give blood with him. They used their combined loot—$250—to buy beer and burgers and threw a party that evening at the ramshackle house they rented a few blocks away.

THE FIRST THING that dampened Tyler's enthusiasm for working at Theranos was seeing the inside of an Edison. During his internship the previous summer, he hadn't been allowed near one, so his anticipation was high when a Chinese scientist named Ran Hu showed him one of the machines with its black-and-white case removed. Standing next to Tyler was Aruna Ayer, his supervisor. Aruna was just as curious as he was: in her previous role as head of the protein engineering group, she had never seen an Edison either. As Ran did a quick demonstration, Tyler and Aruna weren't sure what to think. The device seemed to consist of nothing more than a pipette fastened to a robotic arm that moved back and forth on a gantry. Both had envisioned some sort of sophisticated microfluidic system. But this seemed like something a middle-schooler could build in his garage.

Trying to keep an open mind, Aruna asked, "Ran, do you think this is cool?"

In a tone that implied she did not, Ran replied, "I'll let you decide for yourself."

When its case was back on, the Edison did sport a touchscreen software interface, but even that was a letdown. You had to pound on the screen's icons to get it to work. Tyler and some other members of the group joked that Steve Jobs would have rolled over in his grave if he had seen one of them. Tyler felt a wave of disappointment wash over him but beat it back by telling himself that the 4S, the next-generation device he had heard was in the works, was probably much more intricate.

Soon, there were other things that began to trouble Tyler. One type of experiment he and Erika were tasked with doing involved retesting blood samples on the Edisons over and over to measure how much their results varied. The data collected were used to calculate each Edison blood test's coefficient of variation, or CV. A test is generally considered precise if its CV is less than 10 percent. To Tyler's dismay, data runs that didn't achieve low enough CVs were simply discarded and the experiments repeated until the desired number was reached. It was as if you flipped a coin enough times to get ten heads in a row and then declared that the coin always returned heads. Even within the "good" data runs, Tyler and Erika noticed that some values were deemed outliers and deleted. When Erika asked the group's more senior scientists how they defined an outlier, no one could give her a straight answer. Erika and Tyler might be young and inexperienced, but they both knew that cherry-picking data wasn't good science. Nor were they the only ones who had concerns about these practices. Aruna, whom Tyler liked and respected, also disapproved of them and so did Michael Humbert, a jovial German scientist Tyler had befriended.

One of the validation experiments Tyler helped with involved a test to detect syphilis. Some tests measure the concentration of a substance in the blood, such as cholesterol, to determine whether it is too high. Others, like the syphilis test, provide a yes-or-no answer about whether a patient has a particular disease or not. The accuracy of those tests is gauged by their sensitivity—the measure of how often they correctly label someone with the disease as positive. Over a period of several days, Tyler and several colleagues tested 247 blood samples on Edisons, 66 of which were known to be positive for the disease. During the first run, the devices correctly detected only 65 percent of the positive samples. During the second run, they correctly detected 80 percent of them. Yet, in its validation report, Theranos stated that its syphilis test had a sensitivity of 95 percent.

Erika and Tyler thought Theranos was also being misleading about the accuracy of other Edison tests, such as a test to measure vitamin D. When a blood sample would be tested on an analyzer made by the Italian company DiaSorin, it might show a vitamin D concentration of 20 nanograms per milliliter, which was considered adequate for a healthy patient. But when Erika tested the same sample on the Edison, the result would be 10 or 12 nanograms per milliliter—a value that signified a vitamin D deficiency. The Edison's vitamin D test was nonetheless cleared for use in the clinical lab on live patient samples, as were two Edison thyroid hormone tests and a test to measure PSA, the prostate cancer marker.

———

IN NOVEMBER 2013, Erika was moved from the immunoassay group to the clinical lab and assigned to Normandy, the room downstairs with the lab's Edison machines. During the Thanksgiving holiday, a patient order came in from the Walgreens store in Palo Alto for a vitamin D test. As she had been trained to do, Erika ran a quality-control check on the Edison devices before testing the patient sample.

Quality-control checks are a basic safeguard against inaccurate results and are at the heart of the way laboratories operate. They involve testing a sample of preserved blood plasma that has an already-known concentration of an analyte and seeing if the lab's test for that analyte matches the known value. If the result obtained is two standard deviations higher or lower than the known value, the quality-control check is usually deemed to have failed.

The first quality-control check Erika ran failed, so she ran a second one. That one failed too. Erika was unsure what to do. The lab's higher-ups were on vacation, so she emailed an emergency help line the company had set up. Sam Anekal, Suraj Saksena, and Daniel Young responded to her email with various suggestions, but nothing they proposed worked. After a while, an employee named Uyen Do from the research-and-development side came down and took a look at the quality-control readings.

Under the protocol Sunny and Daniel had established, the way Theranos generated a result from the Edisons was unorthodox to say the least. First, the little finger-stick samples were diluted with the Tecan liquid handler and split into three parts. Then the three diluted parts were tested on three different Edisons. Each device had two pipette tips that dropped down into the diluted blood, generating two values. So together, the three devices produced six values. The final result was obtained by taking the median of those six values.

Following this protocol, Erika had tested two quality-control samples across three devices, generating six values during each run for a total of twelve values. Without bothering to explain her rationale to Erika, Do deleted two of those twelve values, declaring them outliers. She then went ahead and tested the patient sample and sent out a result.

This wasn't how you were supposed to handle repeat quality-control failures. Normally, two such failures in a row would have been cause to take the devices off-line and recalibrate them. Moreover, Do wasn't even authorized to be in the clinical lab. Unlike Erika, she didn't have a CLS license and had no standing to process patient samples. The episode left Erika shaken.

LESS THAN a week later, Alan Beam was chatting nervously in Jurassic Park, the upstairs lab, with a female inspector from the Laboratory Field Services division of the California Department of Public Health. The Theranos lab's CLIA certificate was nearly two years old and up for renewal, which required the lab to pass an inspection. The federal Medicare agency outsourced these types of routine inspections to state inspectors.

Sunny had let it be known that no employee was to enter or exit Normandy during the inspection. The stairs that led to the downstairs room were hidden behind a door that could only be opened with a key card. Alan and other members of the lab interpreted the directive as a clear signal that Sunny didn't want the inspector to inquire about what was behind the door. The inspector spent several hours in the upstairs part of the lab and found some relatively minor problems that Alan pledged to correct promptly. Then she was gone—unaware that she had missed the part of the lab that contained the company's proprietary devices. Alan didn't know whether to be relieved or angry. Had he just helped hoodwink a regulator? Why was he being put in this position?

In the days following the inspection, Sunny ordered a switch from regular venous draws to fingerstick draws for dozens of the blood tests Theranos was offering in Walgreens stores, not just the four performed on the Edisons. That meant that the system Daniel Young and Sam Gong had jury-rigged with the Siemens ADVIAs would now be used on regular patients. It didn't take long for problems to surface.

Elizabeth and Sunny had decided to make Phoenix their main launch market, drawn by Arizona's pro-business reputation and its large number of uninsured patients, who they believed would be especially receptive to the low prices Theranos offered. So, in addition to its one Palo Alto location, the company had just opened two wellness centers in Walgreens stores in the Phoenix area, with plans for several dozen more. Elizabeth planned to open a second lab in Phoenix, but for now the fingerstick samples collected in the Arizona stores were being FedExed back to Palo Alto for testing. The arrangement was far from ideal: the nanotainers were shipped in coolers, but the coolers heated up when they sat baking in the sun for hours on the airport tarmac. This caused the blood in the little tubes to clot.

Just as had been the case before the launch when they were still testing employee samples, Alan was also encountering issues with potassium results. The blood in the nanotainers was often pink, a telltale sign of hemolysis, and the potassium results the diluted samples were producing were consistently too high. Some were so high that the only way they could have been accurate was if the patients were dead. The problem got so bad that Alan implemented a rule that no potassium result above a certain threshold could be released to a patient. He pleaded with Elizabeth to pull the potassium test from the Theranos menu. Instead, she sent Daniel Young to try to fix the assay.

IN EARLY 2014, Tyler Shultz was moved from the immunoassay group to the production team, which operated downstairs in Normandy. This put him back near Erika and other colleagues from the clinical lab who were processing patient samples on the Edisons and the modified Siemens ADVIAs. There were no physical barriers between the two groups, so Tyler could hear the chatter among the lab associates. Tyler learned from Erika and others that the Edisons were frequently flunking quality-control checks and that Sunny was pressuring lab personnel to ignore the failures and to test patient samples on the devices anyway.

As he debated what to do, he got a call from his grandfather. George said he was throwing Elizabeth a thirtieth birthday party and he wanted his grandson to come and play a tune for her. Tyler had been playing the guitar since high school and liked to compose his own songs. During his travels the previous summer, he'd played in pubs and on street corners around Ireland. Tyler tried to get out of it by invoking work: his shift on the production team was from 3:00 p.m. to 1:00 a.m., overlapping with the evening party. But George insisted. He'd already made a seating chart and placed his grandson between Channing Robertson and Elizabeth at the dinner table. And he was sure Elizabeth wouldn't mind if Tyler missed work to celebrate her birthday. She wanted him there, he said.

A few days later, Tyler found himself mingling with other guests in the living room of George's home, a big light-blue shingled house perched on a hill next to the Stanford campus. George's second wife, Charlotte, was playing host to the festivities. Elizabeth's parents had flown in for the occasion and her younger brother, Christian, was there too. So were Channing Robertson and Theranos board member Bill Perry, who had served as secretary of defense in the Clinton administration.

At his grandfather's urging, Tyler played the song he'd hastily composed. He tried not to cringe as he sang its cheesy lyrics, which borrowed from Theranos's "one tiny drop changes everything" slogan. To his horror, he had to play it again a little while later because Henry Kissinger arrived late and everybody thought that he too should hear it. When Tyler was finished, Kissinger, who like George Shultz was in his early nineties, recited a limerick he'd written for the birthday girl. The scene had a surreal quality to it: they were all sitting in a circle in the Shultzes' living room and Elizabeth was in the middle, reveling in the attention. It was as though she were the queen and they were her court, kissing her ring. As awkward as the evening was, it made Tyler feel like he was on friendly enough terms with Elizabeth to speak to her candidly about his concerns. Shortly after the party, he sent her an email asking if they could meet.

Elizabeth invited him to her office. Their meeting was brief, but he had time to raise a few of the issues that bothered him. One of them was the representations Theranos made about the precision of its blood tests: the company claimed that its tests had coefficients of variation of less than 10 percent, but the CVs in many of its validation reports were much higher, he told her. Elizabeth acted surprised and said she didn't think Theranos had made such a claim. She suggested they look at its website together and called it up on her big iMac. A part of the site titled "Our Technology" did prominently advertise a coefficient of variation of less than 10 percent with a catchy green-and-white circular logo, but Elizabeth noted that the smaller print above it specified that the claim only covered Theranos's vitamin D test.

Tyler conceded her point and made a mental note to check the vitamin D validation data. He then brought up the fact that his own CV calculations often didn't match those he found in the validation reports. By his count, the percentages in the reports were lower than they should be. In other words, Theranos was exaggerating the precision of its blood tests.

"That doesn't sound right," Elizabeth said. She suggested he go speak with Daniel Young. Daniel would be able to walk him through how Theranos performed its data analyses and clear up any confusion. Over the following weeks, Tyler met with Daniel Young twice. Talking to Daniel could be frustrating. He had a long forehead accentuated by a receding hairline that suggested a big, powerful brain. But it was impossible to know what went on inside that brain. His eyes, behind their wire-rim glasses, never betrayed any emotion.

During their first meeting, Daniel calmly explained why Tyler's CV calculations were wrong: Tyler was taking into account the six values, or "replicates," generated during each Edison test instead of just the median of those six values. The final result Theranos reported to a patient was the median, so only that number was relevant to CV calculations, he said.

Daniel may have technically been correct, but Tyler had put his finger on a central weakness of the Edison device: its pipette tips were terribly imprecise. Generating six measurements during each test and then selecting the median was a way to correct for that imprecision. If the tips had been reliable in the first place, there would have been no need for such contortions.

The conversation shifted to the syphilis test and what Tyler felt was its overstated sensitivity. Again, Daniel had a ready explanation: some of the

Edisons' syphilis results had fallen into an equivocal zone. Results in that zone hadn't been included in the sensitivity calculation. Tyler remained dubious. There didn't seem to be any predefined criteria for this so-called equivocal zone. It could be widened at will until the sensitivity reached whatever number the company wanted. In the case of the syphilis test, it was so wide that more samples had been deemed equivocal than the Edisons had correctly identified as positive. Tyler asked Daniel if he thought Theranos's syphilis test was truly the most accurate syphilis test on the market, as the company claimed. Daniel replied that Theranos had never claimed to have the most accurate tests.

After Tyler got back to his desk, he googled the two recent articles that had been published in the press about Theranos and emailed them to Daniel. One of them was Elizabeth's **Wall Street Journal** interview, which stated that Theranos's tests were "more accurate than the conventional methods" and called that improved accuracy a scientific advance. When they met again a few days later, Daniel allowed that the statements in the **Journal** piece were too sweeping but argued that they had been made by the writer, not by Elizabeth herself. Tyler found this argument a little too convenient. Surely the writer hadn't made up these claims on his own; he must have heard them from Elizabeth. A faint smile briefly crossed Daniel's lips.

"Well, sometimes Elizabeth exaggerates in an interview setting," he said.

There was something else that was bothering Tyler—something he'd just gotten wind of from Erika—and he decided to bring that up too. All clinical laboratories must submit three times a year to something called "proficiency testing," an exercise designed to ferret out labs whose testing isn't accurate. Accredited bodies like the College of American Pathologists send laboratories samples of preserved blood plasma and ask them to test them for various analytes.

During its first two years of operation, the Theranos lab had always tested proficiency-testing samples on commercial analyzers. But since it was now using the Edisons for some patient tests, Alan Beam and his new lab codirector had been curious to see how the devices fared in the exercise. Beam and the new codirector, Mark Pandori, had ordered Erika and other lab associates to split the proficiency-testing samples and run one part on the Edisons and the other part on the lab's Siemens and DiaSorin analyzers for comparison. The Edison results had differed markedly from the Siemens and DiaSorin ones, especially for vitamin D.

When Sunny had learned of their little experiment, he'd hit the roof. Not only had he put an immediate end to it, he had made them report only the Siemens and DiaSorin results. There was a lot of chatter in the lab that the Edison results should have been the ones reported. Tyler had looked up the CLIA regulations and they seemed to bear that out: they stated that proficiency-testing samples must be tested and analyzed "in the same manner" as patient specimens "using the laboratory's routine methods." Theranos tested patient samples for vitamin D, PSA, and the two thyroid hormones on the Edisons, so it followed that the proficiency-testing results for those four analytes should have come from the Edisons.

Tyler told Daniel he didn't see how what Theranos had done could be legal. Daniel's response followed a tortuous logic. He said a laboratory's proficiency-testing results were assessed by comparing them to its peers' results, which wasn't possible in Theranos's case because its technology was unique and had no peer group. As a result, the only way to do an apples-to-apples comparison was by using the same conventional methods as other laboratories. Besides, proficiency-testing rules were extremely complicated, he argued. Tyler could rest assured that no laws had been broken. Tyler didn't buy it.

AT 9:16 A.M. on Monday, March 31, 2014, the email Tyler had been waiting for all weekend landed in his Yahoo in-box—or rather in the inbox of Colin Ramirez, an alias he had made up to remain anonymous. The email was from Stephanie

Shulman, director of the Clinical Laboratory Evaluation Program at the New York State Department of Health. She was responding to a query Tyler had submitted the previous Friday under the cover of his new fictional identity.

Tyler had reached out to the New York health department because it ran one of the proficiency-testing programs Theranos had participated in. He still suspected that the way the company conducted proficiency testing was improper and he wanted an expert opinion. After exchanging a few emails with Shulman, Tyler had his answer. In response to a description he gave her of Theranos's practices, she wrote back that they amounted to "a form of PT cheating" and were "in violation of the state and federal requirements." Shulman gave Tyler two options: he could give her the name of the offending laboratory, or he could file an anonymous complaint with New York State's Laboratory Investigative Unit. He chose to do the latter.

Armed with the knowledge that he was correct about his proficiency-testing suspicions, Tyler went to see his grandfather. They sat down together in the dining room of George's big house, and Tyler tried to explain to the former secretary of state the concepts of precision, sensitivity, quality control, and proficiency testing and to show him why he thought Theranos's approach to each was lacking. He also revealed that Theranos was using its proprietary device for only a handful of the more than two hundred blood tests it advertised on its website. And that before samples could even be processed on the device, they had to be diluted with a third-party machine six feet long and two and a half feet wide that cost tens of thousands of dollars.

George took it all in quizzically. Tyler could tell he wasn't getting through to him, but he needed him to know as both his grandfather and a member of the company's board of directors that he could no longer be a party to what was going on. He told him he planned to quit. George asked him to hold off and to give Elizabeth another chance to address everything. Tyler agreed to do so and tried to set up another meeting with Elizabeth, but her rising public profile made her very busy. She asked him to send her an email with his concerns instead. So he went ahead and typed up a long note that summarized his conversations with Daniel Young and explained why he'd found most of Daniel's answers unconvincing. He even included charts and validation data to illustrate his various points. He closed with

> I am sorry if this email sounds attacking in any way, I do not intend it to be, I just feel a responsibility to you to tell you what I see so we can work towards solutions. I am invested in this company's long-term vision, and am worried that some of our current practices will prevent us from reaching our bigger goals.

Tyler didn't hear anything back for several days. When the response finally arrived, it didn't come from Elizabeth. It came from Sunny. And it was withering. In a point-by-point rebuttal that was longer than Tyler's original email, Sunny belittled everything from his grasp of statistics to his knowledge of laboratory science. The overall message was that Tyler was too junior and green to understand what he was talking about. Sunny's tone throughout was dripping with venom, but he reserved his sharpest words for the questions Tyler had raised about proficiency testing:

> That reckless comment and accusation about the integrity of our company, its leadership and its core team members based on absolute ignorance is so insulting to me that had any other person made these statements, we would have held them accountable in the strongest way. The only reason I have taken so much time away from work to address this personally is because you are Mr. Shultz's grandson . . .
>
> I have now spent an extraordinary amount of time postponing critical business matters to investigate your assertions—the only email on this topic I want to see from you going forward is an apology that I'll pass on to other people including Daniel here.

Tyler decided it was time to resign. He replied to Sunny with a one-sentence email giving his two weeks' notice and offering to leave earlier if he wished him to. A few hours later, Mona, the head of HR, summoned him to her office and informed him that the company had decided he should leave that day. She made him sign some new nondisclosure forms and told him security would escort him out of the building. But no one from security was available to come get him, so Tyler saw himself out.

He hadn't even made it to his car when his cell phone rang. It was his mother and she sounded frantic.

"Stop whatever you're about to do!" she implored.

Tyler told her it was too late. He had already resigned and signed his exit papers.

"That's not what I mean. I just got off the phone with your grandfather. He said Elizabeth called him and told him that if you insist on carrying out your vendetta against her, you will lose."

Tyler was dumbfounded. Elizabeth was threatening him through his family, using his grandfather to deliver the message. He felt a surge of anger. After hanging up with his mother, he headed over to the Hoover Institution.

George Shultz's secretary showed him to his grandfather's corner office on the second floor of the Herbert Hoover Memorial Building. A lifetime's worth of books lined the shelves. Tyler was still unnerved by Elizabeth's threat but calmly explained to George what had happened. He showed him his email to Elizabeth and Sunny's blistering reply. George asked his secretary to make photocopies of them and to put them in his office safe.

Tyler thought he might be getting through this time, but he wasn't sure. The old man was hard to read. His years as a senior member of the president's cabinet, facing down threats like the Soviet Union at the height of the Cold War, had made him a cipher. He absorbed information but rarely volunteered any. They agreed to meet again for dinner that evening at his grandfather's house. As they parted, George told Tyler, "They're trying to convince me that you're stupid. They can't convince me that you're stupid. They can, however, convince me that you're wrong and in this case I do believe that you're wrong."

ERIKA KNEW THAT Tyler had quit and asked herself if she should do the same. Things in the lab had gotten out of control. In addition to the four original Edison tests, the assay validation team had cleared a hepatitis C test on the Edisons for clinical use. Giving patients inaccurate vitamin D results was one thing, but the stakes got a lot higher when you were testing for infectious diseases.

A patient order for a hepatitis C test had come in and Erika had refused to run the sample on the Edisons. When Mark Pandori had asked her to come talk to him about it, she'd broken down in tears in his office. Erika and Mark had a good relationship and Erika trusted him. Ever since he'd arrived a few months earlier, Mark had tried to do the right thing, including with proficiency testing.

Erika told Mark the reagents for the hepatitis C test were expired, the Edisons hadn't been recalibrated in a while, and she simply didn't trust the devices. So they had devised a plan to run patient samples on commercially available hepatitis kits called OraQuick HCV. That had worked for a while, but then the lab had run out of them. When they'd tried to place an order for a new batch, Sunny had lost his temper and threatened to block it.

Then, that very afternoon, at about the same time Tyler had gotten his mother's frantic call, Sunny had summoned her to his office. He had gone through Tyler's emails and figured out that Erika was the one who had sent him the proficiency-testing results. Their conversation had started out cordially enough, but Sunny had berated her when she'd brought up the quality-control failures in the lab. His parting words had been, "You need to tell me if you want to work here or not."

When her shift was over, Erika went to meet up with Tyler. He suggested she accompany him to his grandfather's house for dinner. If George saw that his grandson wasn't the only employee with misgivings about the way Theranos operated, he might come around. Erika agreed that it was worth a try.

When they got there, however, it quickly became apparent to Tyler that his grandfather's allegiance to Theranos had strengthened in the intervening hours. As the Shultzes' household staff waited on them, Tyler and Erika ran through the list of their concerns, but only George's wife, Charlotte, seemed receptive to what they were saying. She kept asking them in a shocked tone of voice to repeat various parts of their story.

George, on the other hand, was unmoved. Tyler had noticed how much he doted on Elizabeth. His relationship with her seemed closer than their own. Tyler also knew that his grandfather was passionate about science. Scientific progress would make the world a better place and save it from such perils as pandemics and climate change, he often told his grandson. This passion seemed to make him unable to let go of the promise of Theranos.

George said a top surgeon in New York had told him the company was going to revolutionize the field of surgery and this was someone his good friend Henry Kissinger considered to be the smartest man alive. And according to Elizabeth, Theranos's devices were already being used in medevac helicopters and hospital operating rooms, so they must be working.

Tyler and Erika tried to tell him that couldn't possibly be true given that the devices were barely working within the walls of Theranos. But it was clear they weren't making any headway. George urged them to put the company behind them and to move on with their lives. They both had bright futures ahead, he told them. They left the dinner frustrated, with little choice but to follow his advice.

The next morning, Erika quit too. She wrote up a short resignation letter and gave it to Mark Pandori to pass on to Elizabeth and Sunny. It said she disagreed with running patient samples on the Edisons and that she didn't think she and the company shared "the same standards in patient care and quality." After taking a look at it, Mark gave it back to her and recommended she leave quietly without making waves.

Erika thought about it for a moment and decided he was probably right. She folded the letter back up and put it in her backpack. But while processing Erika's resignation in her office a few minutes later, Mona asked if she had taken anything from the company. To show she hadn't, Erika opened her backpack and showed her its contents. Mona spotted the letter inside and confiscated it. She made Erika sign a new confidentiality agreement and warned her against writing anything about Theranos on Facebook, LinkedIn, or any other forum.

"We have ways of tracking that," she said. "We'll see it if you post anything anywhere."

| SEVENTEEN |

Fame

Richard and Joe Fuisz sat warily across from David Boies and one of his partners at a table in the lobby lounge of San Jose's Fairmont hotel. It was a Sunday evening in the middle of March and the usually bustling lounge's two grand pianos were quiet, allowing the four men to speak without raising their voices. Boies, looking relaxed and dapper in a navy blazer and his signature black sneakers, had called the meeting to discuss settling the litigation that had pitted the Fuiszes against Theranos for the past two and a half years.

Initially determined to fight the lawsuit to the bitter end, Richard and Joe were tired and battered. The trial had started a few days earlier at the federal courthouse down the street and the extent to which they were outgunned had fully dawned on them. Unhappy with their lawyers and their mounting legal costs, they had gone "pro se" several months earlier. What had seemed then like a reasonable decision now looked foolish: Joe, a patent attorney who had never tried a case, was no match for the country's best litigator and his army of associates.

The death of Ian Gibbons had also been a big setback. It briefly looked like they might be able to make up for it by calling his widow, Rochelle, as a witness. After Richard managed to make contact with her, Rochelle told him that Elizabeth had tried to bully Ian into not testifying and that Ian thought she was dishonest. But the judge overseeing the case had denied the Fuiszes' late motion to call Rochelle to the stand.

More damaging, though, had been Richard Fuisz's own courtroom testimony two days earlier. Boies had caught him in a series of pointless lies that, while they did nothing to prove Theranos's theft allegations, had undermined his credibility. One of them was Fuisz's contention that he still practiced medicine and treated patients—a claim his own wife had refuted in her deposition. For no discernible reason other than pride, Fuisz had refused to back away from it even after Boies confronted him with her testimony. In his rambling opening argument, Fuisz had also stated that his patent had nothing to do with Theranos, which was absurd on its face given that his patent application mentioned the company by name and quoted from its website.

Joe had watched his father's disastrous performance on the stand with growing alarm. His dad had once been an amazing pitchman in business settings because he was a terrific schmoozer and improviser, but that off-the-cuff, loose-with-the-facts approach didn't work when you were being questioned under oath by a legal ace ready to pounce on any inconsistency. It didn't help that, at seventy-four years old, Richard's memory was beginning to slip.

Joe feared his brother John's upcoming testimony might turn into another liability. Boies knew John had a bad temper and would no doubt find ways to press his buttons in front of the jury. He had already brought up the fact that John had threatened Elizabeth during his deposition.

When he added it all up in his mind, Joe knew they were in trouble. And with a courtroom defeat looking like a very real possibility, he was haunted by a terrifying thought: What if they not only lost, but the judge made them cover Theranos's legal expenses? He shivered to think how much money their opponent was spending on the case. He worried it might be enough to bankrupt him and his father both. They had already spent more than $2 million on their defense.

Boies had come to their powwow with Mike Underhill, one of the Boies Schiller attorneys running point on the litigation. Underhill, a very tall and gangly man, broke the ice by asking Richard Fuisz if he had really grown up on a farm (the answer was yes). That led to Fuisz and Boies discussing raising cattle, which Boies had some experience with from the ranch he owned in Napa Valley. When the conversation eventually turned to the matter at hand, Underhill said both sides would be better off settling the case. If, however, the Fuiszes remained intent on pressing forward with the trial, they should know that matters would be revealed that would destroy John Fuisz. Underhill didn't specify what nor did he say this menacingly. He made it sound as though he liked John and it would pain him to see him get hurt. There was some irony to Underhill's threat to air dirty laundry about John. The two of them had once been colleagues at McDermott Will & Emery and had shared a secretary. Underhill had left McDermott not long after John had made a sexual harassment complaint against him on the secretary's behalf to the firm's human resources department. (Underhill denies any untoward behavior and says his departure from McDermott to join Boies Schiller was already in motion.)

The prospect of new damaging information coming out about his brother added yet another worry to Joe's long list, but the truth was that he and his father had come to the meeting ready to settle. It didn't take long for an agreement to take shape: the Fuiszes would withdraw their patent in exchange for Theranos withdrawing its suit. No money would change hands; each party would remain responsible for its legal costs. It amounted to a complete capitulation on the Fuiszes' part. Elizabeth had won.

Boies insisted they draft the agreement right then and there. He wrote the terms down on a piece of paper and passed it to Joe, who made a few modifications. Underhill then took it upstairs to have it typed up. As they waited for Underhill to return, Richard Fuisz complained once more that Elizabeth's theft accusation was false. Playing the part of the magnanimous victor, Boies allowed that that might be the case but he had a client to answer to.

Fuisz asked Boies if he could do something for John. His son's reputation had been unjustly sullied, he said. Underhill had previously raised with Joe the notion that Boies Schiller could refer patent work to John if he signed a release promising not to sue Elizabeth or the firm. Boies repeated that offer. He would need to wait six months for things to quiet down, but then he could start sending work John's way. He suggested they get John on the phone to talk it over.

Fuisz dialed John's number in Washington and passed his cell phone to Boies. As it turned out, John was in no mood to make nice. He had been looking forward to testifying in court. He saw it as his chance to clear his name. Now this settlement would prevent him from doing that. He angrily told Boies there was no way he would ever sign a release unless Theranos issued a public statement exonerating him. Richard and Joe could see the conversation wasn't going well: Boies was holding the phone several inches from his ear and wincing as John shouted on the other end of the line. After a few minutes, Boies passed the phone back to Fuisz. Their little side deal was dead.

But the main agreement stood. When Underhill came back with the printed settlement, Richard and Joe read it and signed it. Afterward, Richard Fuisz looked utterly defeated. The proud and pugnacious former CIA agent broke down and sobbed.

THE NEXT MORNING, Fuisz jotted down a note on a paper pad from the hotel and, when he got to the courthouse, asked Boies to pass it on to Elizabeth. It read:

> Dear Elizabeth,
> This matter is resolved now. I wish great success for you and health and happiness for your parents. We all can be wrong. Life is like that. Please know that in fact none of the 612 patent came from any of your provisionals. It derived from my brain only.
>
> Best wishes,
> Richard Fuisz

Back in Washington, the settlement didn't sit well with John Fuisz. He was mad at everyone, including his father and brother, for agreeing to a deal that gave Theranos everything it wanted before he'd had a chance to tell his side of the story in court. In his pique, John emailed a young reporter named Julia Love, who had been covering the case for American Lawyer Media, and told her about the quid pro quo Boies had sought the night before, making it sound like an attempt to bribe him. He also vowed to sue Boies and to add his father and brother to the suit as defendants. He then forwarded the email to Underhill and to Richard and Joe, letting them know that anything they sent his way would be forwarded to the media.

Underhill responded angrily a few hours later, leaving the reporter off his reply but copying his boss. He denied any attempt to bribe John and warned him that Boies Schiller would hold him responsible if he continued to make such claims. In case the message wasn't clear, Boies himself chimed in from his iPad a few minutes later: **Those who the gods would destroy, they first make mad.**

JULIA LOVE'S ARTICLE about the settlement in **Litigation Daily,** ALM's newsletter, caught the eye of Roger Parloff, **Fortune** magazine's legal correspondent. Parloff, who had once practiced law as a white-collar criminal defense attorney in

Manhattan before becoming a journalist, was always on the lookout for legal sagas to write about.

This particular case struck him as strange and, in his experience, strange cases usually made for good yarns. Why had Boies, arguably the country's most famous lawyer with his pick of high-profile cases to choose from, handled this obscure patent trial himself instead of delegating it to a more junior associate? Then there was the fact that John Fuisz, an attorney who was the son of one of the defendants and brother of the other, was publicly threatening to sue both the plaintiff and Boies for making false accusations.

From his office in the Time & Life Building in Midtown Manhattan, Parloff picked up the phone and called Dawn Schneider, Boies's longtime public-relations representative. Parloff's call was perfect timing from Schneider's perspective. She had just talked to an ebullient Boies about the case and decided she should try to get him some press about it. She offered to come brief the **Fortune** writer in person. The Boies Schiller offices at Fifty-First Street and Lexington Avenue were just four avenue blocks away.

As she walked across Midtown, it occurred to Schneider that Boies's victory in the Fuisz case was a good story but the far better story was Theranos and its brilliant young founder. She had never met Elizabeth, but she'd been hearing Boies rave about her for several years. This was an opportunity to get David's protégée national attention just as her company prepared to expand across the country. By the time she got to the **Fortune** offices on Avenue of the Americas, Schneider had changed her pitch.

Parloff listened intrigued. He hadn't seen the **Wall Street Journal** article from the previous fall so he had never heard of Theranos but, according to Schneider, that was precisely the point. It was like writing about Apple or Google in their early days before they became Silicon Valley icons and entered the collective consciousness.

"Roger, this is the greatest company you've never heard of," she said. "Think of it as an old-school **Fortune** cover."

A few weeks later, Parloff flew out to Palo Alto to meet Elizabeth. Over the course of several days, he interviewed her for a combined seven hours. After getting over his initial shock at her deep voice, he found her smart and engaging. When they broached topics other than blood testing, she was unassuming, almost naïve. But when their conversations shifted to Theranos, she became steely and intense. She was also very controlling with information. She dangled a scoop: Theranos had raised more than $400 million from investors at a valuation of $9 billion, making it one of the most valuable startups in Silicon Valley. And she showed Parloff the miniLab (though she didn't refer to it by any name). But she wouldn't let the magazine take photos of it and she didn't want Parloff to use the words "device" or "machine" to describe it. She preferred "analyzer."

Leaving those quirks aside, what Elizabeth told Parloff she'd achieved seemed genuinely innovative and impressive. As she and Sunny had stated to Partner Fund, she told him the Theranos analyzer could perform as many as seventy different blood tests from one tiny finger-stick draw and she led him to believe that the more than two hundred tests on its menu were all finger-stick tests done with proprietary technology. Since he didn't have the expertise to vet her scientific claims, Parloff interviewed the prominent members of her board of directors and effectively relied on them as character witnesses. He talked to Shultz, Perry, Kissinger, Nunn, Mattis, and to two new directors: Richard Kovacevich, the former CEO of the giant bank Wells Fargo, and former Senate majority leader Bill Frist. Before going into politics, Frist had been a heart and lung transplant surgeon. All of them vouched for Elizabeth emphatically. Shultz and Mattis were particularly effusive.

"Everywhere you look with this young lady, there's a purity of motivation," Shultz told him. "I mean she really is trying to make the world better, and this is her way of doing it."

Mattis went out of his way to praise her integrity. "She has probably one of the most mature and well-honed sense of ethics—personal ethics, managerial ethics, business ethics, medical ethics that I've ever heard articulated," the retired general gushed.

Parloff didn't end up using those quotes in his article, but the ringing endorsements he heard in interview after interview from the luminaries on Theranos's board gave him confidence that Elizabeth was the real deal. He also liked to think of himself as a pretty good judge of character. After all, he'd dealt with his share of dishonest people over the years, having worked in a prison during law school and later writing at length about such fraudsters as the carpet-cleaning entrepreneur Barry Minkow and the lawyer Marc Dreier, both of whom went to prison for masterminding Ponzi schemes. Sure, Elizabeth had a secretive streak when it came to discussing certain specifics about her company, but he found her for the most part to be genuine and sincere. Since his angle was no longer the patent case, he didn't bother to reach out to the Fuiszes.

WHEN PARLOFF'S COVER STORY was published in the June 12, 2014, issue of **Fortune,** it vaulted Elizabeth to instant stardom. Her **Journal** interview had gotten some notice and there had also been a piece in **Wired,** but there was nothing like a magazine cover to grab people's attention. Especially when that cover featured an attractive young woman wearing a black turtleneck, dark mascara around her piercing blue eyes, and bright red lipstick next to the catchy headline "THIS CEO IS OUT FOR BLOOD."

The story disclosed Theranos's valuation for the first time as well as the fact that Elizabeth owned more than half of the company. There was also the now-familiar comparison to Steve Jobs and Bill Gates. This time it came not from George Shultz but from her old Stanford professor Channing Robertson. (Had Parloff read Robertson's testimony in the Fuisz trial, he would have learned that Theranos was paying him $500,000 a year, ostensibly as a consultant.) Parloff also included a passage about Elizabeth's phobia of needles—a detail that would be repeated over and over in the ensuing flurry of coverage his story unleashed and become central to her myth.

When the editors at **Forbes** saw the **Fortune** article, they immediately assigned reporters to confirm the company's valuation and the size of Elizabeth's ownership stake and ran a story about her in their next issue. Under the headline "Bloody Amazing," the article pronounced her "the youngest woman to become a self-made billionaire." Two months later, she graced one of the covers of the magazine's annual Forbes 400 issue on the richest people in America. More fawning stories followed in **USA Today, Inc., Fast Company,** and **Glamour,** along with segments on NPR, Fox Business, CNBC, CNN, and CBS News. With the explosion of media coverage came invitations to numerous conferences and a cascade of accolades. Elizabeth became the youngest person to win the Horatio Alger Award. **Time** magazine named her one of the one hundred most influential people in the world. President Obama appointed her a U.S. ambassador for global entrepreneurship, and Harvard Medical School invited her to join its prestigious board of fellows.

As much as she courted the attention, Elizabeth's sudden fame wasn't entirely her doing. Her emergence tapped into the public's hunger to see a female entrepreneur break through in a technology world dominated by men. Women like Yahoo's Marissa Mayer and Facebook's Sheryl Sandberg had achieved a measure of renown in Silicon Valley, but they hadn't created their own companies from scratch. In Elizabeth Holmes, the Valley had its first female billionaire tech founder.

Still, there was something unusual in the way Elizabeth embraced the limelight. She behaved more like a movie star than an entrepreneur, basking in the public adulation she was receiving. Each week brought a new media interview or conference appearance. Other well-known startup founders gave interviews and made public appearances too but with nowhere near the same frequency. The image of the reclusive, ascetic young woman Parloff had been sold on had overnight given way to that of the ubiquitous celebrity.

Elizabeth was also quick to embrace the trappings

of fame. The Theranos security team grew to twenty people. Two bodyguards now drove her around in a black Audi A8 sedan. Their code name for her was "Eagle One." (Sunny was "Eagle Two.") The Audi had no license plates—another nod to Steve Jobs, who used to lease a new Mercedes every six months to avoid having plates. Elizabeth also had a personal chef who prepared her salads and green vegetable juices made of cucumber, parsley, kale, spinach, lettuce, and celery. And when she had to fly somewhere, it was in a private Gulfstream jet.

PART OF WHAT made Elizabeth's persona so compelling was her heartwarming message about using Theranos's convenient blood tests to catch diseases early so that, as she put it in interview after interview, no one would have to say goodbye to loved ones too soon. In September 2014, three months after the **Fortune** cover story, she made that message more poignant during a speech at the TEDMED conference in San Francisco by adding a personal dimension to it: for the first time, she told the story in public of her uncle who had died of cancer—the same story Tyler Shultz had found so inspiring when he'd started working at Theranos.

It was true that Elizabeth's uncle, Ron Dietz, had died eighteen months earlier from skin cancer that had metastasized and spread to his brain. But what she omitted to disclose was that she had never been close to him. To family members who knew the reality of their relationship, using his death to promote her company felt phony and exploitative. Of course, no one in the audience at San Francisco's Palace of Fine Arts knew this. Most of the one thousand spectators in attendance found her performance mesmerizing.

Clad all in black, she strode solemnly around the stage as she spoke, like a preacher giving a sermon. In a stunt that made for brilliant theater, she pulled a nanotainer out of her jacket pocket midway through and held it up to illustrate how little blood Theranos's tests required. Calling the fear of needles "one of the basic human fears, up there with the fear of spiders and the fear of heights," she then told other touching anecdotes. One was about a little girl who got stuck repeatedly with a syringe by a hospital nurse who couldn't find her vein. Another was about cancer patients whose spirits were broken by all the blood they had to give as part of their treatments.

One of the people watching from a seat halfway up the auditorium was Patrick O'Neill, whom Elizabeth had hired away from TBWA\Chiat\Day and appointed Theranos's chief creative officer. Patrick had become instrumental in honing Elizabeth's image and raising her profile. He had helped her prepare for the conference and before that had worked with **Fortune**'s photographer on the magazine's cover shoot. To Patrick, making Elizabeth the face of Theranos made perfect sense. She was the company's most powerful marketing tool. Her story was intoxicating. Everyone wanted to believe in it, including the numerous young girls who were sending her letters and emails. It wasn't a cynical calculus on his part: Patrick was one of her biggest believers. He had no knowledge of the shenanigans in the lab and didn't pretend to understand the science of blood testing. As far as he was concerned, the fairy tale was real.

Before he became a full-time employee, Elizabeth had hung inspirational quotes in little frames around the old Facebook building. One of them was from Michael Jordan: "I've missed more than 9,000 shots in my career. I've lost almost 300 games. 26 times, I've been trusted to take the game winning shot and missed. I've failed over and over and over again in my life. And that is why I succeed." Another was from Theodore Roosevelt: "Far and away the best prize that life has to offer is the chance to work hard at work worth doing."

Patrick suggested they make them a more integral part of the workplace by painting them in black on the building's white walls. Elizabeth liked the idea. She also loved a new quote he suggested. It was from Yoda in **Star Wars:** "Do or do not. There is no try." She had it painted in huge capital letters in the building's entrance.

To accommodate its swelling ranks, which now totaled more than five hundred, Theranos was planning to move to a new location it had leased from Stanford a few blocks away on Page Mill Road. It was the site of an old printing plant that had been demolished. Patrick was put in charge of the new building's interior and hired the South African architect Clive Wilkinson, who had designed the converted Chiat\Day warehouse in L.A., for the job.

The central motif of the design was once again the sacred geometry of the circle. Desks were arranged in large circular patterns rippling out from circular glass conference rooms in the center. The carpeting followed the same circular patterns. In the building's lobby, interlocking rings of brass were embedded in the floor's terrazzo tiles to form the Flower of Life symbol. Elizabeth's new corner office was designed to look like the Oval Office. Patrick ordered a custom-made desk that was as deep as the president's at its center but had rounded edges. In front of it, he arranged two sofas and two armchairs around a table, replicating the White House layout. At Elizabeth's insistence, the office's big windows were made of bulletproof glass.

Patrick wasn't just Elizabeth's style and décor consultant. He also spearheaded a big marketing push Theranos was making in Arizona, where its wellness centers had expanded to forty Walgreens stores. He hired Errol Morris, an Academy Award–winning documentary filmmaker who moonlighted as a producer and director of commercials, to make video ads the company ran on TV stations in the Phoenix area and on its website and YouTube channel. One of the spots was a close-up of Elizabeth in her customary black turtleneck staring into the camera and talking about what she called people's "basic human right" to access their own health information through blood tests. Her eyes looked so big and she spoke so slowly and deliberately that the video had a hypnotic quality to it.

Another spot featured patients complaining about how much they hated big needles and then acting pleased at the painlessness of the Theranos experience as they got their fingers pricked. Patrick thought it was powerful and arranged for it to run during shows with high female viewership, like the ABC drama **Scandal,** because research had shown that mothers were households' medical decision makers. But the ad had to be pulled a couple of weeks after it started airing because a local doctor complained that some of his patients had gone to Walgreens stores expecting a finger-stick draw only to be told their tests required a needle after all. Patrick was disappointed but didn't raise a fuss about it because he knew this was a sensitive subject. Several months earlier, he had asked Sunny what proportion of Theranos tests were performed with finger-stick draws versus regular venous ones. Sunny had refused to give him a straight answer and had abruptly changed the subject.

| EIGHTEEN |

The Hippocratic Oath

Alan Beam, Theranos's laboratory director, was late to the party.

A white tent had been set up on basketball courts next to the old Facebook building, which the company was in the process of vacating. Music was blaring from big outdoor speakers and lights projected images of giant pink spiders onto the makeshift dance floor. The grass field behind the tent was decorated with pumpkins and bales of hay. As he breathed in the cooling evening air of Palo Alto's Indian summer, Alan scanned the costumed crowd and caught sight of Elizabeth. She was wearing a long velvet dress with gold trim and a big upright collar, and her blond hair was done up in an elaborate bun. The irony of her Queen Elizabeth attire wasn't lost on him. With a net worth that **Forbes** had just estimated at $4.5 billion in its October 20, 2014, issue, she had become Silicon Valley royalty.

Elizabeth loved to throw company parties. And none more so than the one she organized every year for Halloween. It was a Theranos tradition for which no expense was spared. The company's senior executives all played along. Sunny was dressed as an Arab sheik. Daniel Young was Walter White, the high school chemistry teacher turned drug dealer from the show **Breaking Bad**. Christian Holmes and the Frat Pack were characters from Quentin Tarantino's **Kill Bill** movies.

Ordinarily stiff and aloof at the office, Elizabeth liked to let loose on these occasions. At last year's party, she'd jumped up and down in a bouncy house like an excited child. This year, the bouncy house was replaced by an inflatable boxing ring. While employees wearing sumo suits and oversized boxing gloves wobbled around in it, Elizabeth was delighting in the costume of an engineer disguised as a giant neutrophil.

Alan was supposed to be a zombie and he felt like one. In retrospect, leaving his tranquil post in Pittsburgh to come work at Theranos had been like crossing into his own strange version of the **Twilight Zone**. For his first few months as laboratory director, he'd clung to the belief that the company was going to transform lab testing with its technology. But the past year's events had shattered any illusion of that. He now felt like a pawn in a dangerous game being played with patients, investors, and regulators. At one point, he'd had to talk Sunny and Elizabeth out of running HIV tests on diluted finger-stick samples. Unreliable potassium and cholesterol results were bad enough. False HIV results would have been disastrous.

His codirector, Mark Pandori, had quit after just five months on the job. The trigger had been a request he'd made that Elizabeth check in with them before making representations to the press about Theranos's testing capabilities. Sunny had summarily rejected it, prompting Mark to hand in his resignation that very day. Another member of the lab had been so troubled by some of the company's practices that she told Alan she couldn't sleep at night. She too had resigned.

Alan was reaching his own breaking point. A few weeks earlier, he'd started forwarding dozens of work emails to his personal Gmail account. He knew forwarding the emails was a risky move because the company monitored everything, but he wanted to keep a record of the concerns he'd repeatedly raised with Sunny and Elizabeth. He'd gone a step further two days earlier and called a law firm in Washington, D.C., that specialized in representing corporate whistleblowers, but the person who answered the phone was a "client services specialist." He opted to keep the reason for his call vague, wanting to speak only to an attorney. He did send them one of his email exchanges with Sunny, but he worried it was hard to understand without additional context and a good understanding of how clinical laboratories operate.

It was also hard to prove everything. The company kept things so compartmentalized. Why wasn't he being shown quality-control data anymore? How could a lab director, the person who was supposed to vouch for the accuracy of the test results delivered to doctors and patients, be denied that information? His other big concern was proficiency testing. After reading up on the CLIA regulations, he'd become convinced that Theranos was gaming the exercise.

"Alaaan!"

Daniel Young had sidled up next to him, interrupting his somber thoughts. As was his habit at these work parties, Daniel was drunk. The alcohol made him uncharacteristically friendly and approachable, but Alan knew better than to share his misgivings. Daniel was part of the inner circle. They made small talk, bantering about Daniel's upper-crust upbringing in Connecticut. As they chatted, the festivities seemed to be winding down. Some colleagues were headed to Antonio's Nut House, a dive bar a few blocks down the street, to have a few more beers. Alan and Daniel tagged along.

When they got to the bar, Alan spotted Curtis Schneider, a scientist on the R&D side of the company, and grabbed a stool beside him. Curtis was one of the smartest people Alan knew at Theranos. He had a Ph.D. in inorganic chemistry and had spent four years as a postdoctoral scholar at Caltech. They talked about fly fishing for a while. It was one of Curtis's favorite hobbies. Then Curtis told Alan about a conference call earlier that day with officials from the FDA. Theranos was trying to get the agency to approve some of its proprietary blood tests. During the call, one of the agency's reviewers voiced a dissenting view about the company's submission but was silenced by his colleagues. Curtis found it odd. There might be nothing to it, Alan thought, but the story added to his mounting unease. He told Curtis about the lab's quality-control data and how it was being kept from him. And he confided something else: the company was cheating on proficiency testing. In case Curtis hadn't registered the implication of what he'd just said, he spelled it out: Theranos was breaking the law.

When he looked up, Alan saw that Young was staring at them from across the bar. His face was white as a ghost's.

THREE WEEKS LATER, Alan was sitting in his new office in Newark when he got a call from Christian Holmes. Most of the company had moved to the new building on Page Mill Road in Palo Alto, but not the clinical laboratory. The lab had moved across San Francisco Bay to Theranos's sprawling

Newark facility, where it planned to one day manufacture thousands of miniLabs.

Christian wanted Alan to handle yet another doctor's complaint. Alan had fielded dozens of them since the company had gone live with its tests the previous fall. Time and time again, he'd been asked to convince physicians that blood-test results he had no confidence in were sound and accurate. He decided he couldn't do it anymore. His conscience wouldn't allow him to.

He told Christian no and emailed Sunny and Elizabeth to inform them that he was resigning and to ask them to immediately take his name off the lab's CLIA license. Elizabeth replied that she was deeply disappointed. He agreed to delay his official departure by another month to give Theranos time to find a new lab director. For the first two weeks of his notice period, Alan went on vacation. He rode his motorcycle to Los Angeles to see his brother for a few days and then flew to New York to spend Thanksgiving with his parents. When he returned in mid-December, he headed to the new Palo Alto headquarters to discuss a transition plan with Sunny.

Sunny came down with Mona to meet him in the lobby of the new building. They ushered him into a room off the reception area and informed him he was being terminated early. Sunny slid what looked like a legal document across the table toward him.

Alan read the bold heading at the top: "AFFIDAVIT OF ALAN BEAM."

It stated, under penalty of perjury under the laws of California, that he promised to never disclose any proprietary or confidential information learned during his employment at the company. And it included this line: "I do not have any electronic or hard copy information relating to Theranos in my possession in any location including personal email accounts, any personal laptops or desktops, trash/deleted folders, USB drives, home, car, or any other location."

Before Alan had time to finish reading, he heard Sunny say in an icy tone, "We know you sent yourself a bunch of work emails. You have to let Mona access your Gmail account so that she can go through them and delete them."

Alan refused. He told Sunny the company had no right to invade his privacy and he wouldn't sign any more documents.

Sunny's face reddened. His volcanic temper was building. Shaking his head in disgust, he turned to Mona and said, "Can you believe this guy?"

He turned back to Alan. Contempt oozing from his voice, he offered to hire him an attorney to expedite matters.

The notion that a lawyer in the pay of Theranos would adequately defend his interests in a dispute with the company struck Alan as absurd. He declined the offer and announced that he wanted to leave. Mona gave him his backpack, which he'd insisted she retrieve from the lab. In return, she asked for his company phone and laptop. He handed them over after quickly resetting the phone to factory settings to wipe it of its contents. Then he walked out.

Over the next few days, the messages piled up in his voicemail box. Some were from Sunny, others from Mona. They all said the same thing, in increasingly threatening tones: he needed to come back to the office, let Mona delete the emails from his personal email account, and sign the affidavit. Or else the company would sue him.

Alan realized that they weren't going to stop. He needed a lawyer. Contacts with that Washington firm had gone nowhere. He needed someone local he could consult in person. He called the first listing that came up in a Google search: a medical malpractice and personal injury attorney in San Francisco. She agreed to represent him after he paid her a ten-thousand-dollar retainer.

As his new lawyer saw it, Alan didn't have much of a choice. Theranos could make a case that his actions did breach his confidential obligations. And even if it failed to do so, it could tie him up in court for months, if not years. This was one of the most valuable private companies in Silicon Valley, one of the fabled unicorns. Its financial resources were virtually limitless. The litigation could bankrupt him. Did he really want to take that risk?

His lawyer was getting pressure from a Boies

Schiller partner who was representing Theranos and she was clearly intimidated. She urged Alan to delete the emails and to sign the affidavit. She told him she would send Theranos a preservation order instructing it to keep the originals. There was no assurance the company would honor it, but that was the best they could do, she said.

That evening, Alan glumly sat down at his computer in his apartment in Santa Clara and logged into his Gmail account. One by one, he erased the emails. By the time he was done, he'd counted 175 of them.

IT HAD BEEN nine months since Richard Fuisz had settled with Theranos and agreed to withdraw his patent, but he was still consumed with the case. For the first few weeks after the settlement, he'd been nearly catatonic. His wife, Lorraine, had had to call his son Joe to find out what had happened because he refused to talk about it.

During the litigation, Fuisz had found a friendly ear in Phyllis Gardner, a longtime friend who was a professor at Stanford's medical school. Phyllis and her husband, Andrew Perlman, had briefly been involved with Theranos during its infancy because Elizabeth had consulted Phyllis about her original patch idea when she dropped out of Stanford. After telling her she didn't think her concept was remotely feasible, Phyllis had referred Elizabeth to

Andrew, a veteran biotech industry executive. Andrew had agreed to serve on a short-lived Theranos advisory board that Elizabeth had dissolved after just a few months.

The decade-old episode had left Phyllis skeptical that Elizabeth, who had no medical or scientific training to speak of and a clear tendency not to listen to people who were older and more experienced, had really gone on to develop groundbreaking blood-testing technology. Her suspicions had deepened when Andrew had chatted with a Siemens sales representative during a flight and learned that Theranos was a major purchaser of Siemens diagnostic equipment.

Fuisz too had become doubtful that Theranos could really do what it claimed. During a visit to Palo Alto for pretrial motions in the fall of 2013, he'd called the local Walgreens and asked if he could have a creatinine finger-stick test done there. His doctors had recently diagnosed him with aldosteronism, a hormonal disorder that causes high blood pressure, and wanted him to monitor his creatinine levels for any sign of kidney damage. Creatinine is a common blood test, but the woman who answered the phone told him the wellness center didn't offer it without a special approval from Theranos's CEO. When he added that to the company's intense secrecy and to the fact that it had actively discouraged Ian Gibbons from testifying before he died, he smelled a rat.

Fuisz had introduced Phyllis to Ian's widow, Rochelle, and the two women had bonded over their distrust of Elizabeth. Together, the three of them formed a little band of Theranos skeptics. The problem was that no one else seemed to share their doubts.

That would change when, in its December 15, 2014, issue, **The New Yorker** published a profile of Elizabeth. In many ways, it was just a longer version of the **Fortune** story that had rocketed her to fame six months earlier. The difference this time was that someone knowledgeable about blood testing read it and was immediately dubious.

That someone was Adam Clapper, a practicing pathologist in Columbia, Missouri, who in his spare time wrote a blog about the industry called **Pathology Blawg**. To Clapper, it all sounded too good to be true, especially Theranos's supposed ability to run dozens of tests on just a drop of blood pricked from a finger.

The **New Yorker** article did strike some skeptical notes. It included quotes from a senior scientist at Quest who said he didn't think finger-stick blood tests could be reliable, and it noted Theranos's lack of published, peer-reviewed data. Among the arguments she marshaled to rebut the latter point, Elizabeth cited a paper she had coauthored in a medical journal called **Hematology Reports**. Clapper had never heard of **Hematology Reports** before, so he looked into it. He learned that it was an online-only publication based in Italy that charged scientists who wanted to publish in it a five-hundred-dollar fee. He then looked up the paper Holmes had co-authored and was shocked to see that it included data for just one blood test from a grand total of six patients.

In a post on his blog about the **New Yorker** story, Clapper pointed out the medical journal's obscurity and the flimsiness of the study and declared himself a skeptic "until I see evidence Theranos can deliver what it says it can deliver in terms of diagnostic accuracy." **Pathology Blawg** didn't exactly have a big readership, but Joe Fuisz came across the post in a Google search and brought it to his father's attention. Richard Fuisz immediately got in touch with Clapper and told him he was onto something. He put him in contact with Phyllis and Rochelle and urged him to listen to what they had to say. Clapper was intrigued by what the three of them told him, especially by the story of Ian Gibbons's death. But it all sounded too circumstantial to go beyond what he'd already written. What he needed was some sort of proof, he told Fuisz.

Fuisz was frustrated. What would it take for people to listen to him and finally see through Elizabeth Holmes?

While checking his emails a few days later, Fuisz saw a notification from LinkedIn alerting him that someone new had looked up his profile on the site. The viewer's name—Alan Beam—didn't ring a bell but his job title got Fuisz's attention: laboratory director at Theranos. Fuisz sent Beam a message through the site's InMail feature asking if they could talk on the phone. He thought the odds of getting a response were very low, but it was worth a try. He was in Malibu taking photos with his old Leica camera the next day when a short reply from Beam appeared in his iPhone in-box. He was willing to talk and he included his cell-phone number. Fuisz drove his black Mercedes E-Class sedan back to Beverly Hills and, when he was just a few blocks from home, dialed the number.

The voice he heard on the other end of the line sounded terrified. "Dr. Fuisz, the reason I'm willing to talk to you is you're a physician," Beam said. "You and I took the Hippocratic Oath, which is to first do no harm. Theranos is putting people in harm's way." Alan proceeded to tell Fuisz about a litany of problems in the Theranos lab. Fuisz pulled into his driveway and quickly got out of his car. As soon as he was inside his house, he grabbed a notepad he'd brought back from a stay at a hotel in Paris called Le Meurice and started taking notes. Alan was speaking so fast that he was having trouble keeping up with what he was saying. He jotted down:

LIED TO CLIA people & cheated
ROLL OUT DISASTER
Finger stick not accurate—using venipuncture

Transporting Arizona to Palo Alto
Using Siemens equip.
Ethical breaches
False thyroid results
K results all over map
False pregnancy errors
Told Eliz not ready but insisted proceed

Fuisz asked Alan to talk to Joe and to Phyllis. He wanted them to hear it for themselves from the horse's mouth. Alan agreed to call them and more or less repeated to each of them what he had told Fuisz. But that was all he was willing to do. He wouldn't talk to anyone else. Boies Schiller lawyers had been hounding him, he said, and he couldn't afford to be sued like Fuisz had been. Although he sympathized with Alan's predicament, Fuisz couldn't just leave it at that. He got back in touch with Clapper and told him about the new connection he had made and what he'd learned. This was the proof he'd been asking for, he told him.

Clapper agreed that this changed everything. The story now had legs. But he decided he couldn't take it on himself. For one thing, he couldn't shoulder the legal liability of going up against a $9 billion Silicon Valley company with a litigious track record that was represented by David Boies. For another, he was just an amateur blogger. He didn't have the journalistic know-how to tackle something like this. Not to mention the fact that he had a full-time medical practice to tend to. This, he thought, was a job for an investigative reporter. In the three years since he'd launched **Pathology Blawg,** Clapper had spoken to several about lab-industry abuses. There was one in particular who came to mind. He worked for the **Wall Street Journal**.

| NINETEEN |

The Tip

It was the second Monday in February and I was sitting at my messy desk in the **Wall Street Journal**'s Midtown Manhattan newsroom casting about for a new story to sink my teeth into. I'd recently finished work on a year-long investigation of Medicare fraud and had no idea what to do next. After sixteen years at the **Journal,** this was something I still hadn't mastered: the art of swiftly and efficiently transitioning from one investigative project to the next.

My phone rang. It was Adam from **Pathology Blawg**. I'd sought his help eight months earlier when I was trying to understand the complexities of laboratory billing for one of my stories in the Medicare series. He'd patiently explained to me what lab procedures certain billing codes corresponded to—knowledge I'd later used to expose a scam at a big operator of cancer treatment centers.

Adam told me he'd stumbled across what he thought could be a big story. People often come to journalists with tips. Nine times out of ten, they don't pan out, but I always took the time to listen. You never knew. Besides, at this particular moment, I was like a dog without a bone. I needed a new bone to chew on.

Adam asked if I'd read a recent feature in **The New Yorker** about a Silicon Valley prodigy named Elizabeth Holmes and her company, Theranos. As it turned out, I had. I subscribed to the magazine and often read it on the subway to and from work.

Now that he mentioned it, there were some things I'd read in that article that I'd found suspect. The lack of any peer-reviewed data to back up the company's scientific claims was one of them. I'd reported about health-care issues for the better part of a decade and couldn't think of any serious advances in medicine that hadn't been subject to peer review. I'd also been struck by a brief description Holmes had given of the way her secret blood-testing devices worked: "A chemistry is performed so that a chemical reaction occurs and generates a signal from the chemical interaction with the sample, which is translated into a result, which is then reviewed by certified laboratory personnel."

Those sounded like the words of a high school

chemistry student, not a sophisticated laboratory scientist. The **New Yorker** writer had called the description "comically vague."

When I stopped to think about it, I found it hard to believe that a college dropout with just two semesters of chemical engineering courses under her belt had pioneered cutting-edge new science. Sure, Mark Zuckerberg had learned to code on his father's computer when he was ten, but medicine was different: it wasn't something you could teach yourself in the basement of your house. You needed years of formal training and decades of research to add value. There was a reason many Nobel laureates in medicine were in their sixties when their achievements were recognized.

Adam said that he'd had a similar reaction to the **New Yorker** piece and that a group of people had contacted him after he'd posted a skeptical item on his blog about it. He was cryptic about their identities and their connection to Theranos at first, but he said they had information about the company I'd want to hear. He said he'd check with them to see if they were willing to talk to me.

In the meantime, I did some preliminary research on Theranos and came across the **Journal**'s editorial-page piece from seventeen months earlier. I hadn't seen it when it was published. This added an interesting wrinkle, I thought: my newspaper had played a role in Holmes's meteoric rise by being the first mainstream media organization to publicize her supposed achievements. It made for an awkward situation, but I wasn't too worried about it. There was a firewall between the **Journal**'s editorial and newsroom staffs. If it turned out that I found some skeletons in Holmes's closet, it wouldn't be the first time the two sides of the paper had contradicted each other.

Two weeks after our initial conversation, Adam put me in touch with Richard and Joe Fuisz, Phyllis Gardner, and Rochelle Gibbons. It was disappointing at first to hear that the Fuiszes had been involved in litigation with Theranos. Even if they insisted they'd been wrongly accused, the lawsuit gave them a big ax to grind and made them useless as sources.

But my ears pricked up when I heard that they had talked to Theranos's just-departed laboratory director and that he was alleging some sort of wrongdoing at the company. I also found the story of Ian Gibbons tragic and was intrigued by the fact that Rochelle said he'd confided to her on several occasions that the Theranos technology wasn't working. It was the type of thing that would have been dismissed as hearsay in court, but it seemed credible enough to merit a closer look. In order to take this any further, though, what I needed to do next was clear: I needed to talk to Alan Beam.

THE FIRST HALF dozen times I dialed Alan's number, I got his voicemail. I didn't leave a message and instead resolved to just keep trying him. On the afternoon of Thursday, February 26, 2015, a voice with an accent I couldn't quite place finally answered the phone. After ascertaining that it was in fact Alan, I introduced myself and told him I understood he had recently left Theranos with concerns about the way the company was operating.

I could sense he was very nervous, but he also seemed to want to unburden himself. He told me he would speak to me only if I promised to keep his identity confidential. Theranos's lawyers had been harassing him and he was certain the company would sue him if it found out he was talking to a reporter. I agreed to grant him anonymity. It wasn't a hard decision. Without him, all I had were secondhand sources and informed speculation. If he wouldn't talk, there would be no story.

With the ground rules for our conversation established, Alan let down his guard and we talked for more than an hour. One of the first things he said was that what Ian had told Rochelle was true: the Theranos devices didn't work. They were called Edisons, he said, and were error prone. They constantly failed quality control. Furthermore, Theranos used them for only a small number of tests. It performed most of its tests on commercially available instruments and diluted the blood samples.

It took me a while to understand the dilution part. Why would they do that and why was it bad?

I asked. Alan explained that it was to make up for the fact that the Edison could only do a category of tests known as immunoassays. Theranos didn't want people to know its technology was limited, so it had contrived a way of running small finger-stick samples on conventional machines. This involved diluting the finger-stick samples to make them bigger. The problem, he said, was that when you diluted the samples, you lowered the concentration of analytes in the blood to a level the conventional machines could no longer measure accurately.

He said he had tried to delay the launch of Theranos's blood tests in Walgreens stores and had warned Holmes that the lab's sodium and potassium results were completely unreliable. According to Theranos's tests, perfectly healthy patients had levels of potassium in their blood that were off the charts. He used the word "crazy" to describe the results. I was barely getting my head around these revelations when Alan mentioned something called proficiency testing. He was adamant that Theranos was breaking federal proficiency-testing rules. He even referred me to the relevant section of the Code of Federal Regulations: 42 CFR, part 493. I wrote it down in my notebook and told myself to look it up later.

Alan also said that Holmes was evangelical about revolutionizing blood testing but that her knowledge base in science and medicine was poor, confirming my instincts. He said she wasn't the one running Theranos day-to-day. A man named Sunny Balwani was. Alan didn't mince his words about Balwani: he was a dishonest bully who managed through intimidation. Then he dropped another bombshell: Holmes and Balwani were romantically involved. I knew from reading the **New Yorker** and **Fortune** articles and from browsing the Theranos website that Balwani was the company's president and chief operating officer. If what Alan was saying was true, this added a new twist: Silicon Valley's first female billionaire tech founder was sleeping with her number-two executive, who was nearly twenty years her senior.

It was sloppy corporate governance, but then again this was a private company. There were no rules against that sort of thing in Silicon Valley's private startup world. What I found more interesting was the fact that Holmes seemed to be hiding the relationship from her board. Why else would the **New Yorker** article have portrayed her as single, with Henry Kissinger telling the magazine that he and his wife had tried to fix her up on dates? If Holmes wasn't forthright with her board about her relationship with Balwani, then what else might she be keeping from it?

Alan said he had raised his concerns about proficiency testing and the reliability of Theranos's test results with Holmes and Balwani a number of times in person and by email. But Balwani would always either rebuff him or put him off, making sure to copy a Theranos lawyer on their email exchanges and to write, "Consider this attorney-client confidential."

As the laboratory director whose name had been on the Theranos lab's CLIA license, Alan was worried that he would be held personally responsible if there was ever a government investigation. To protect himself, he told me he'd forwarded dozens of his email exchanges with Balwani to his personal email account. But Theranos had found that out and threatened to sue him for breaching his confidentiality agreement.

What worried him even more than any personal liability he might face was the potential harm patients were being exposed to. He described the two nightmare scenarios false blood-test results could lead to. A false positive might cause a patient to have an unnecessary medical procedure. But a false negative was worse: a patient with a serious condition that went undiagnosed could die.

I hung up the phone feeling the familiar rush I got whenever I made a big reporting breakthrough and had to remind myself that this was just the first step in a long process. There was still a lot to understand and, above all, the story would require corroboration. There was no way the paper would take it with just one anonymous source, however good that source might be.

THE NEXT TIME Alan and I talked, I was standing in Brooklyn's Prospect Park trying to stay warm while keeping a loose eye on my two boys, ages nine and eleven, as they horsed around with one of their friends. It was the last Saturday in what would go down in the record books as New York City's coldest February in eighty-one years.

I had texted Alan after our first conversation to ask if he could think of former colleagues who might corroborate what he'd told me. He'd sent seven names, and I'd made contact with two of them. Both had been extremely nervous and had only agreed to talk on deep background. One of them, a former Theranos CLS, wouldn't say much, but what she did say gave me confidence that I was on the right track: she told me she had been very troubled by what was going on at the company and concerned for patient safety. She'd resigned because she wasn't comfortable having her name continue to appear on test results. The other was a former technical supervisor in the lab who'd said that Theranos operated under a culture of secrecy and fear.

I told Alan that I felt like I was beginning to make progress, which he seemed pleased to hear. I asked him whether he had kept the emails he'd forwarded to his personal Gmail account. My heart sank when he responded that his lawyer had made him delete them to comply with the affidavit the company made him sign. Documentary evidence was the gold standard for these types of stories. This would make my job much more difficult. I tried not to betray my disappointment.

Our conversation shifted to proficiency testing. Alan explained how Theranos was gaming it and he told me which commercial analyzers it used for the majority of its blood tests. Both were made by Siemens, confirming what Andrew Perlman, Phyllis Gardner's husband, had heard from a Siemens sales representative during a flight. He revealed something else that hadn't come up in our first call: Theranos's lab was divided into two parts. One contained the commercial analyzers and the other the Edison devices. During her inspection of the lab, a state inspector had been shown only the part with the commercial analyzers. Alan felt she'd been deceived.

He also mentioned that Theranos was working on a newer-generation device code-named 4S that was supposed to supplant the Edison and do a broader variety of blood tests, but it didn't work at all and was never deployed in the lab. Diluting finger-stick samples and running them on Siemens machines was supposed to be a temporary solution, but it had become a permanent one because the 4S had turned into a fiasco.

It was all beginning to make sense: Holmes and her company had overpromised and then cut corners when they couldn't deliver. It was one thing to do that with software or a smartphone app, but doing it with a medical product that people relied on to make important health decisions was unconscionable. Toward the end of this second phone conversation, Alan mentioned something else I found of interest: George Shultz, the former secretary of state who was a Theranos board member, had a grandson named Tyler who had worked at the company. Alan wasn't sure why Tyler had left but he didn't think it was on good terms. I was jotting things down in the Notes app of my iPhone and added Tyler's name as another potential source.

OVER THE NEXT few weeks, I made some more progress but I also encountered some complications. In my quest to corroborate what Alan was telling me, I contacted more than twenty current and former Theranos employees. Many didn't reply to my calls and emails. The few that I managed to get on the phone told me they had signed ironclad confidentiality agreements and didn't want to risk being sued for violating them.

One former high-ranking lab employee did agree to talk to me but only off the record. This was an important journalistic distinction: Alan and the other two former employees had agreed to speak to me on deep background, which meant I could use what they told me while keeping their identities confidential. Off the record meant I couldn't make any use of the information. The conversation was nonetheless helpful because this source confirmed a lot of what Alan had told me, giving me the confidence to forge on. He summed up what was going on at the company with an analogy: "The way Theranos is operating is like trying to build a bus while you're driving the bus. Someone is going to get killed."

A few days later, Alan got back in touch with some good news. I had asked him to call the Washington, D.C., whistleblower law firm he'd reached out to in the fall to see if he could retrieve the email exchange with Balwani he had sent to it. The firm had just complied with his request. Alan forwarded the exchange to me. It was a chain of eighteen emails about proficiency testing between Sunny Balwani, Daniel Young, Mark Pandori, and Alan. It showed Balwani angrily admonishing Alan and Mark Pandori for running the proficiency-testing samples on the Edison and reluctantly acknowledging that the device had "failed" the test. Moreover, it left no doubt that Holmes knew about the incident: she was copied on most of the emails.

This was another step forward, but it was soon followed by a step backward. In late March, Alan got cold feet. He stood by everything he'd told me, but he no longer wanted to be involved with the story going forward. He couldn't stomach the risks anymore. Talking to me gave him palpitations and distracted him from his new job, he said. I tried to get him to change his mind but he was resolute, so I decided to give him some space and hoped he would eventually come around.

Although it was a big setback, I was slowly making headway on other fronts. Wanting a neutral opinion from a lab expert about Theranos's dilution of blood samples and the way it conducted proficiency testing, I called Timothy Hamill, vice chairman of the University of California, San Francisco's Department of Laboratory Medicine. Tim confirmed to me that both practices were highly questionable. He also explained the pitfalls of using blood pricked from a finger. Unlike venous blood drawn from the arm, capillary blood was polluted by fluids from tissues and cells that interfered with tests and made measurements less accurate. "I'd be less surprised if they told us they were time travelers who came back from the twenty-seventh century than if they told us they cracked that nut," he said.

Before his change of heart, Alan had mentioned a nurse in Arizona named Carmen Washington who worked at a clinic owned by Walgreens and had complained about Theranos's blood tests. After trying to track her down for several weeks, I finally got her on the phone. She told me three of her patients had received questionable results from the company. One was a sixteen-year-old girl with a sky-high potassium result that suggested she was at risk of a heart attack. The result hadn't made sense given that she was a teenager and in good health, Carmen said. Two other patients had received results showing abnormally high levels of thyroid-stimulating hormone, or TSH. Carmen had called them back in to the clinic and redrawn their blood. The second time, their results had come back abnormally low. After that, Carmen had lost faith in Theranos's finger-stick tests. These incidents tracked with Alan's claims. TSH was one of the immunoassays Theranos performed on the Edison that had failed proficiency testing.

Carmen Washington's story was helpful, but I soon had something better: another Theranos whistleblower. I had dropped Tyler Shultz a note through LinkedIn's InMail messaging feature after noticing that he had viewed my profile on the site. I figured he must have heard from other former employees that I was poking around. It had been more than a month since I'd made the overture and I was losing hope that he would reply when my phone rang.

It was Tyler and he seemed eager to talk. However, he was extremely worried that Theranos would come after him. He was calling me from a burner phone that couldn't be traced back to him. After I agreed to grant him confidentiality, he told me in broad strokes the story of his eight months at the company.

Tyler's motivation for talking to me was twofold. Like Alan, he was worried about patients getting inaccurate test results. He was also concerned for his grandfather's reputation. Although he felt certain Theranos would eventually be exposed, he wanted to hasten the process to give his grandfather the chance to clear his name. George Shultz was ninety-four and might not be around all that much longer.

"He made it through Watergate and the Iran-Contra scandal with his integrity intact," Tyler told me. "I'm sure he'll get through Theranos if he's still alive to make things right."

On his way out the door, Tyler had printed his email to Holmes and Balwani's response and smuggled them out under his shirt. He also still had the emails he'd exchanged with the New York State Health Department about proficiency testing. This was music to my ears. I asked him to send me everything, which he promptly did.

It was time to head to Palo Alto. But before going, there was somewhere else I wanted to visit first.

I NEEDED TO PROVE that the company was producing inaccurate blood-test results. The only way to do that was to find doctors who had received questionable lab reports and sent their patients to get retested elsewhere. The best place to go looking was Phoenix, where Theranos had expanded to more than forty locations. My first thought had been to pay a visit to Carmen Washington, but she'd left the Walgreens clinic she worked at on the corner of Osborn Road and Central Avenue and didn't have the names of the three patients she'd told me about.

I had another lead, though, after scanning Yelp to see if anyone had complained about a bad experience with Theranos. Sure enough, a woman who appeared to be a doctor and went by "Natalie M." had. Yelp has a feature that allows you to send messages to reviewers, so I sent her a note with my contact information. She called me the next day. Natalie M.'s real name was Nicole Sundene. She was a family practitioner in the Phoenix suburb of Fountain Hills and she was very unhappy with Theranos. The previous fall, she had sent one of her patients to the emergency room because of a frightening lab report from the company only to find out it was a false alarm. I flew to Phoenix to meet Dr. Sundene and her patient. While there, I also planned to drop in unannounced on other physician practices that used Theranos for their lab tests. I'd gotten the names of a half dozen from an industry source.

Dr. Sundene's patient, Maureen Glunz, agreed to meet at a Starbucks near her home. A petite woman in her mid-fifties, she was Exhibit A for one of the two scenarios Alan Beam worried about. The lab report she'd received from Theranos had shown abnormally elevated results for calcium, protein, glucose, and three liver enzymes. Since she had complained of ringing in her ear (later determined to be caused by lack of sleep), Dr. Sundene had worried she might be on the cusp of a stroke and sent her straight to the hospital. Glunz had spent four hours in the emergency room on the eve of Thanksgiving while doctors ran a battery of tests on her, including a CT scan. She'd been discharged after a new set of blood tests performed by the hospital's lab came back normal. That hadn't been the end of it, however. As a precaution, she'd undergone two MRIs during the ensuing week. She said she'd finally stopped worrying when those had come back normal too.

Glunz's case was compelling because it showed both the emotional and the financial toll of a health scare brought on by inaccurate results. As an independent real estate broker, she was self-insured and had a health plan with a high deductible. The ER visit and subsequent MRIs had cost three thousand dollars—a sum she'd had to pay out of her own pocket.

When I met with Dr. Sundene at her office, I learned that Glunz wasn't the only patient whose results she'd found suspect. She told me more than a dozen of her patients had tested suspiciously high for potassium and calcium and she doubted the accuracy of those results as well. She had written Theranos a letter to complain but the company hadn't even acknowledged it.

With Dr. Sundene's help, I decided to conduct a little experiment. She wrote me a lab order and I took it to the Walgreens closest to my hotel the next morning, making sure to fast to ensure the most accurate readings. The Theranos wellness center inside the Walgreens wasn't much to behold: it was a little room not much bigger than a closet with a chair and little bottles of water. Unlike Safeway, the pharmacy chain hadn't spent a fortune remodeling its stores to create upscale clinics. I sat down and waited for a few minutes while the phlebotomist entered my order into a computer and talked to someone on the phone. After she hung up, she asked me to lift up my shirtsleeve and wrapped a tourniquet around my arm. Why no finger stick? I asked. She replied that some of the tests in my order required a venous draw. I wasn't entirely surprised. Alan Beam had explained to me that, of the more than 240 tests Theranos offered on its menu, only about 80 were performed on small fingerstick samples (a dozen on the Edison and another 60 or 70 on the hacked Siemens machines). The rest, he'd said, required what Holmes had likened in media interviews to a medieval torture mechanism: the dreaded hypodermic needle. I now had my confirmation of that. After walking out of the Walgreens, I drove my rental car to a nearby LabCorp site and submitted to another blood draw. Dr. Sundene promised to send me both sets of results when they arrived. Come to think of it, she would get herself tested at both places too, to broaden our comparative sample, she said.

I spent the next few days knocking on the doors of other doctors' offices. At a practice in Scottsdale, I talked to Drs. Adrienne Stewart, Lauren

Beardsley, and Saman Rezaie. Dr. Stewart described a patient of hers who had postponed a long-planned trip to Ireland at the last minute because of a test result from Theranos suggesting she might have deep vein thrombosis, a condition that occurs when a blood clot forms, usually in the legs. People with DVT aren't supposed to fly because of the risk the clot will break loose, travel through the bloodstream, and lodge in the lungs, causing a pulmonary embolism. Dr. Stewart had subsequently set aside the Theranos result when ultrasounds of the patient's legs and a second set of blood-test results from another lab had been normal.

The incident had made her leery when Theranos sent her a lab report for another one of her patients showing an abnormally high TSH value. The patient was already on thyroid medication and the result suggested that her dose needed to be raised. Before she did anything, Dr. Stewart sent the patient to get retested at Sonora Quest, a joint venture of Quest and the hospital system Banner Health. The Sonora Quest result came back normal. Had she trusted the Theranos result and increased the patient's medication dosage, the outcome could have been disastrous, Dr. Stewart said. The patient was pregnant. Increasing her dosage would have made her levels of thyroid hormone too high and put her pregnancy at risk.

I also met with Dr. Gary Betz, a family practitioner in another part of town who had stopped sending his patients to Theranos after a bad experience involving one of them the previous summer. That patient, also a woman, was on medication to reduce her blood pressure. One of the medicine's potential side effects was high potassium, so Dr. Betz monitored her blood regularly. After Theranos reported a near critical potassium value for the patient, a nurse in Dr. Betz's office sent her back there to get tested again to make sure the result was correct. But during the second visit, the phlebotomist made three unsuccessful attempts to draw her blood and then sent her home. Dr. Betz was furious when he found out the next day: if the original result was correct, it was imperative that he get confirmation of it as soon as possible so he could make changes to her treatment. He sent the patient to get retested at Sonora Quest. As it turned out, it was another false alarm: the potassium value Sonora Quest reported that evening was much lower than the Theranos result and well within the normal range. Dr. Betz told me the episode had shattered his trust in Theranos.

As I was wrapping up my trip, I got an email from someone named Matthew Traub. He worked for a public relations firm called DKC and said he represented Theranos. He understood I was working on a story about the company and wanted to know if there was any information he could help me with. The cat was out of the bag, which was just as well. I had planned on contacting the company as soon as I got back to New York. At the **Journal,** we had a cardinal rule called "No surprises." We never went to press with a story without informing the story subject of every single piece of information we had gathered in our reporting and giving them ample time and opportunity to address and rebut everything.

I wrote Traub back to confirm that I had a story in the works. Could he arrange an interview with Holmes and a visit of Theranos's headquarters and laboratory? I asked. I told him I planned on traveling to the San Francisco Bay Area at the beginning of May, which was about two weeks away, and could meet with her then. He said he would check Holmes's schedule and get back to me.

A few days later, I was back at my desk at the **Journal** when a mailroom employee handed me a thick envelope. It was from Dr. Sundene. Inside were our lab reports from Theranos and LabCorp. As I scanned my results, I noticed a number of discrepancies. Theranos had flagged three of my values as abnormally high and one as abnormally low. Yet on LabCorp's report, all four of those values showed up as normal. Meanwhile, LabCorp had flagged both my total cholesterol and LDL cholesterol (otherwise known as bad cholesterol) as high, while the Theranos report described the first as "desirable" and the second as "near optimal."

Those differences were mild compared to a whopper Dr. Sundene had found in her results. According to Theranos, the amount of cortisol in her blood was less than 1 microgram per deciliter. A value that low was usually associated with Addison's disease, a dangerous condition characterized by extreme fatigue and low blood pressure that could result in death if it went untreated. Her LabCorp report, however, showed a cortisol level of 18.8 micrograms per deciliter, which was within the normal range for healthy patients. Dr. Sundene had no doubt which of the two values was the correct one.

WHEN I HEARD back from Traub, he told me Holmes's schedule was too booked up to grant me an interview on such short notice. I decided to fly to San Francisco anyway, to meet Tyler Shultz and Rochelle Gibbons in person. There was also another former Theranos employee who was willing to talk to me if I granted her confidentiality.

The new source met me at a little brewery called Trappist Provisions on College Avenue in Oakland. She was a young woman by the name of Erika Cheung. Like every other former employee I'd spoken to, Erika was very nervous at first. But as I filled her in about how much information I'd already gathered, she visibly relaxed and began telling me what she knew.

As someone who had worked in the Theranos lab, Erika had witnessed the December 2013 lab inspection firsthand. Like Alan, she felt the state inspector had been misled. She told me lab members had been under explicit orders not to enter or exit Normandy during the inspection and that the door leading down to it had been kept locked. She also told me about her friendship with Tyler and about the dinner she'd attended at George Shultz's house the night Tyler had resigned. Like Tyler, she was appalled by the lack of scientific rigor that had gone into validating the assays on the Edisons. She said Theranos should never have gone live testing patient samples. The company routinely ignored quality-control failures and test errors and showed a complete disregard for the well-being of patients, she said. In the end, she had resigned because she was sickened by what she had become a party to, she told me. These were strong words, and it was clear from how distraught Erika was that she meant them.

The next day, I drove to Mountain View, home to Google's headquarters, and met Tyler at a beer garden called Steins. It was early evening and the place was packed with young Silicon Valley professionals enjoying happy hour. We couldn't find seats, so we stood around a wooden beer barrel on the terrace outside and used it as a table. Over a pint of cold ale, Tyler gave me a more detailed account of his time at Theranos, including the frantic call from his mother relaying Holmes's threat the day he resigned and his and Erika's attempts to talk sense into George Shultz that evening. He had tried to follow his parents' advice and to put the whole thing behind him but he'd found himself unable to.

I asked him whether he thought his grandfather was still loyal to Holmes. Yes, there was little doubt in his mind that he was, he replied. When I asked him what made him think that, he revealed a new anecdote. The Shultz family tradition was to celebrate Thanksgiving at the former secretary of state's home. When Tyler, his brother, and his parents had arrived at his grandfather's house that day, they'd come face-to-face with Holmes and her parents. George had invited them too. A mere seven months had passed since Tyler's resignation and the wounds were still fresh, but he had been forced to act as if nothing had happened. The awkward dinner conversation had drifted from California's drought to the bulletproof windows in the new Theranos headquarters. For Tyler, the most excruciating moment had been when Holmes got up and gave a toast expressing her love and appreciation for every member of the Shultz family. He said he'd barely been able to contain himself.

Tyler and Erika were both very young and had been junior employees at Theranos, but I found them credible as sources because so much of what they told me corroborated what Alan Beam had said. I was also impressed by their sense of ethics. They felt strongly that what they had witnessed was wrong and were willing to take the risk of speaking to me to right that wrong.

I next met up with Phyllis Gardner, the Stanford medical school professor Holmes had consulted about her original patch idea when she'd dropped out of college twelve years earlier. Phyllis gave me a tour of the Stanford campus and its surroundings. As we drove around in her car, I was struck by how small and insular Palo Alto was. Phyllis's home was just down the hill from George Shultz's big shingled house, and both were on land owned by Stanford. When Phyllis walked her dog, she sometimes ran into Channing Robertson. The Hoover Institution building where George Shultz and the other Theranos board members had offices was right in the middle of the campus. The new Theranos headquarters on Page Mill Road was less than two miles away on land that was also owned by Stanford. In a strange twist, Phyllis told me the site used to be a **Wall Street Journal** printing plant.

On the last day of my trip, I met Rochelle Gibbons for lunch at Rangoon Ruby, a Burmese restaurant in Palo Alto. It had been two years since Ian had died, but Rochelle was still grieving and struggled to hold back tears. She blamed Theranos for his death and wished he had never worked there. She provided a copy of the doctor's note a Theranos attorney had encouraged Ian to use to avoid being deposed in the Fuisz case. The time stamp on the attorney's email showed that it was sent just a few hours before Ian committed suicide. Rochelle spoke on the record even though she had inherited from her husband Theranos stock options that were potentially worth millions of dollars. She didn't care about the money, she said, and in any case, she didn't believe the shares were really worth anything.

I flew back to New York the next day confident that I'd reached a critical mass in my reporting and that it wouldn't be too long before I could publish. But that was underestimating whom I was up against.

| TWENTY |

The Ambush

The rental house Tyler Shultz shared with five roommates in Los Altos Hills was only a twenty-five-minute drive from his parents' home in Los Gatos, so he tried to have dinner with them about once every other week. Early on the evening of May 27, 2015, Tyler eased his little Toyota Prius C into his parents' garage and entered the house through the kitchen. When he caught sight of his father, he immediately sensed that something was wrong. His face was a mask of worry and panic.

"Have you been speaking to an investigative journalist about Theranos?" his father asked accusingly.

"Yes," Tyler responded.

"Are you kidding me? How stupid could you be? Well, they know."

Tyler learned that his grandfather had just called to say that Theranos was aware he was in contact with a **Wall Street Journal** reporter. If he wanted to get out of what George had described as "a world of trouble," he would need to meet with the company's lawyers the next day to sign something.

Tyler called his grandfather back and asked if the two of them could get together later that night without any lawyers. George said he and Charlotte were out to dinner but should be home by nine and Tyler could come by then. Tyler sat down for a quick meal with his parents, then headed home to think through how he would approach the conversation with his grandfather. His mother and father gave him big hugs as he headed out the door.

When he got home, Tyler called me. From the tone of his voice he seemed a nervous wreck. He asked whether I had disclosed our communications to Theranos. Absolutely not, I replied, telling him I took the promises of confidentiality I made to sources very seriously. We tried to figure out what had happened.

It had been three weeks since we had met at the beer garden in Mountain View. Back in New York, Matthew Traub had continued putting off my requests for an interview with Holmes and had asked me to send him questions instead. I had sent him an email outlining seven main areas I wanted to discuss with Theranos, ranging from Ian Gibbons to proficiency testing.

I forwarded the email to Tyler and he scanned it while we were on the phone. In one section about assay validation, I had included a coefficient of variation for one of the Edison blood tests, not realizing that it was a figure Tyler had calculated himself. There was nothing else in the email that could point to him, so Tyler assumed that was what they had seized on. He seemed to relax. He could easily explain that number away, he said. It could have come from anyone.

Tyler didn't tell me he was about to go see his grandfather, only that Theranos wanted him to come to its offices the next day to meet with its lawyers. I advised him not to go. He no longer worked for the company and was under no obligation to accede to the request. If he went, they would try to smoke him out, I warned him. Tyler said he would think things over. We agreed to touch base again the next day.

TYLER ARRIVED at his grandfather's house at 8:45 p.m. George and Charlotte weren't home yet, so he waited out in the street until he saw their car pull into the driveway. He gave them a few minutes to settle in, then walked into the house. He found them sitting in the living room.

"Have you spoken to any reporters about Theranos?" George asked.

"No," Tyler lied. "I have no idea why they would think that."

"Elizabeth knows you've been talking to the **Wall Street Journal**. She says the reporter used the exact phrasing that's in one of your emails."

Charlotte corrected her husband: "I think she said it was a number."

Was it a number related to proficiency testing? Tyler asked. A lot of people had seen that data, he said. The **Journal** could have gotten it from many other former employees.

"Elizabeth says it could have only come from you," George said sternly.

Tyler stuck to his guns. He said he had no idea how the reporter had gotten his information.

"We're doing this for you," George said. "Elizabeth says your career will be over if the article is published."

Without admitting anything, Tyler tried once more to convince his grandfather that Theranos was misleading him. He went over again all the things he had told him a year earlier, including the fact that the company performed only a small fraction of its blood tests on its proprietary Edison devices. George remained unconvinced. He told Tyler that Theranos had prepared a one-page document for him to sign affirming that he would abide by his confidentiality obligations going forward. The **Wall Street Journal** was going to publish Theranos trade secrets and those trade secrets would become public domain if the company didn't show it had taken action to protect them, he explained. Tyler didn't see why he had to do that but said he'd be willing to consider it if it meant the company would stop bothering him.

"Good, there are two Theranos lawyers upstairs," George said. "Can I go get them?"

Tyler felt blindsided and betrayed. He had specifically asked that they meet without lawyers. But if he tried to duck out now, it would reinforce everyone's suspicion that he had something to hide, so he heard himself say, "Sure."

While George went upstairs, Charlotte told Tyler she was beginning to wonder whether the Theranos "box" was real. "Henry is too," she said, referring to Henry Kissinger, "and he's been saying he wants out."

Before Charlotte had time to say more, a man and a woman appeared and strode aggressively toward Tyler. Their names were Mike Brille and Meredith Dearborn. They were partners at Boies, Schiller & Flexner. Brille told Tyler he had been tasked with finding out who the **Journal**'s sources were and had identified him in about five minutes. He handed him three documents: a temporary restraining order, a notice to appear in court two days later, and a letter stating Theranos had reason to believe Tyler had violated his confidentiality obligations and was prepared to file suit against him.

Tyler again denied having talked to a reporter.

Brille said he knew he was lying and pressed him to admit it, but Tyler stood firm. The lawyer refused to let it rest. He was like an attack dog. He continued to badger Tyler for what felt like an eternity. At one point, Tyler looked at his step-grandmother and asked her if she felt as uncomfortable as he did. Charlotte was glowering at Brille and looked like she was about to give him her right hook.

"This conversation needs to end," Tyler finally said.

George came to his grandson's help. "I know this kid, and this kid doesn't lie. If he says he didn't speak to the reporter, then he didn't speak to the reporter!" he exclaimed. The former secretary of state ushered the two lawyers out of the house. When they were gone, he called Holmes and told her this was not what they had agreed upon. She had sent over a prosecutor rather than someone who was willing to have a civilized conversation. Tyler was ready to go to court the next day, he warned her.

Tyler felt his heart rate accelerate and his hands shake as he watched Charlotte grab the phone from George's hand and heard her say, "Elizabeth, Tyler **did not** say that!"

George got back on the line and a compromise was reached: they would meet again at the house the following morning and Theranos would bring the one-page document they had originally discussed affirming that Tyler would honor his confidentiality obligations. Before hanging up, he implored Holmes to send a different lawyer this time.

THE NEXT MORNING, Tyler arrived at his grandfather's house early and waited in the dining room. He wasn't surprised when it was Brille who showed up again. Holmes was playing his grandfather like a fiddle.

The lawyer had brought along a new set of documents. One of them was an affidavit stating that Tyler had never spoken to any third parties about Theranos and that he pledged to give the names of every current and former employee who he knew had talked to the **Journal**. Brille asked Tyler to sign the affidavit. Tyler refused.

"Tyler isn't a snitch. Finding out who spoke to the **Wall Street Journal** is Theranos's problem, not his," George said.

Brille ignored the former secretary of state and continued pressing Tyler to sign the document and to name the newspaper's sources. Look at things from his perspective, he pleaded: in order to do his job, he needed to get that information from him. But Tyler wouldn't budge.

After the uncomfortable standoff had dragged on long enough, George took Brille into a separate room and came back to talk to Tyler alone. What would it take for him to sign that document? he asked his grandson. Tyler replied that Theranos would have to add a clause promising not to sue him.

George grabbed a pencil and scrawled a line on the affidavit to the effect that Theranos pledged not to sue Tyler Shultz for two years. Tyler wondered for a split second if his grandfather thought he was an idiot.

"That doesn't work for me," he said. "It needs to say they won't ever sue me."

"I'm just trying to come up with something Theranos will agree to," George protested.

But the old man seemed to realize the absurdity of what he'd just proposed. He crossed out the words "two years" and replaced them with "ever." Then he stepped back out of the dining room to go talk to Brille. The two of them returned a few minutes later, with Brille having apparently agreed to Tyler's terms.

The brief interlude had given Tyler time to think, however, and he'd decided he wasn't going to sign anything. One of the other documents Brille had brought that morning was his original Theranos confidentiality agreement. Tyler pretended to read it over while he thought about the best way to say he wouldn't be signing the affidavit. After a long, awkward silence, he settled on how to couch his refusal.

"A Theranos lawyer has drafted this with Theranos's best interests in mind," he said. "I think I need a lawyer to look at it with **my** best interests in mind."

Both his grandfather and Brille looked exasperated. George asked if Tyler would sign the document if his estate attorney, Bob Anders, reviewed it and told him it was OK to do so. Tyler said he would, so George headed upstairs to fax the modified affidavit to Anders. Figuring it would take him a while to get up the stairs and work the fax machine, Tyler went to the kitchen and started going through his grandfather's phone book looking for the estate lawyer's number. He wanted to get to him first. As he was riffling through the pages, Charlotte handed him a piece of paper with the number on it. "Call him," she said.

Tyler placed the call from the backyard. He quickly explained the situation to Anders. Still digesting all the information, the lawyer asked who was representing Theranos. Tyler had the letter threatening to sue him that Brille had given him the night before in his hand. He told Anders it was signed by a "David Boy-zee," mispronouncing the famous lawyer's last name.

"Holy shit! Do you know who that is?"

Boies was one of the most powerful and prominent lawyers in America, Anders explained. This was a serious situation, he said. He recommended that Tyler come see him at his office in San Francisco that afternoon.

Tyler followed his advice and drove to the city. He met with Anders and one of his partners on the seventeenth floor of the Russ Building, a neo-Gothic tower in the Financial District that was once San Francisco's tallest building. After conferring with the two lawyers, Tyler decided not to sign the document. They agreed to communicate that message to Theranos on his behalf, but they would eventually need to refer him to another attorney to avoid a conflict of interest. Their firm, Farella Braun + Martel, also represented Holmes's estate.

When Anders informed Mike Brille that Tyler would not be signing the affidavit, Brille warned that Theranos would have no choice but to sue him. Tyler went home expecting to be summoned to court the next day, but late that evening, Brille sent Anders an email saying Theranos had decided to hold off on a lawsuit to give the two sides more time to work something out. Tyler breathed a sigh of relief when he got the news.

ANDERS REFERRED TYLER to a lawyer named Stephen Taylor, who headed a boutique law firm in San Francisco experienced in handling complex business disputes. During the following weeks, Brille and Taylor exchanged four different drafts of the affidavit.

Tyler tried to be conciliatory in an effort to reach an agreement, acknowledging in the new versions of the document that he had talked to the **Journal**. Theranos gave him the option of saying that he was young and naïve and that the reporter had deceived him, but he declined. He had known exactly what he was doing and youth had nothing to do with it. He hoped he would have acted the exact same way if he had been forty or fifty. To placate Theranos, Tyler did consent to being portrayed as a junior employee whose duties were so menial that he couldn't possibly have known what he was talking about when it came to topics like proficiency testing, assay validation, and lab operations.

But the negotiations stalled over two issues. Theranos still wanted Tyler to name the **Journal**'s other sources, which Tyler steadfastly refused to do. And the company declined to include his parents and heirs in the litigation release it was willing to grant him. As the stalemate dragged on, Boies Schiller resorted to the bare-knuckle tactics it had become notorious for. Brille let it be known that if Tyler didn't sign the affidavit and name the **Journal**'s sources, the firm would make sure to bankrupt his entire family when it took him to court. Tyler also received a tip that he was being surveilled by private investigators. His lawyer tried to make light of it.

"It's not a huge deal," he said. "Just don't go anywhere you're not supposed to be and remember to smile and wave to the man in the bushes outside your house when you leave for work."

One evening, Tyler's parents received a call from his grandfather. George said Holmes had told him that Tyler was responsible for most of the information the **Journal** had and was being completely unreasonable. Tyler's parents sat him down in their kitchen and pleaded with him to sign whatever Theranos wanted him to sign at the next opportunity. Otherwise, they would have to sell their house to pay for his legal costs. It wasn't that simple, Tyler replied, unable to say much more. He badly wanted to explain to them what was going on, but he was under instructions not to discuss the negotiations with Theranos with anyone.

To allow Tyler to update his parents about where things stood, Taylor arranged for them to have their own legal counsel. That way he could communicate with them through attorneys and those conversations would be protected by attorney-client privilege. This led to an incident that rattled both Tyler and his parents. Hours after his parents' new lawyer met with them for the first time, her car was broken into and a briefcase containing her notes from the meeting was stolen. Although it could have been a random act of theft, Tyler couldn't shake his suspicion that Theranos had something to do with it.

I HAD NO IDEA any of this was happening. After Tyler's anxious call the evening he had dinner at his parents' house, I tried to check back in with him. I sent an email to his Colin Ramirez address, which he'd insisted we continue using for his protection, and called him on his burner phone. But my email went unanswered and the phone appeared to be turned off and didn't have a voicemail. I continued to try the email address and the phone for several weeks, to no avail. Tyler had gone dark.

I suspected Theranos was putting the screws to him, but I couldn't confront the company about it since he was a confidential source. I hoped he wouldn't cave under pressure and I took comfort in the fact that he had already sent me his email to Holmes questioning Theranos's practices and the complaint he had filed with New York State. When added to the internal email trail about proficiency testing I had obtained from Alan Beam, it made for a damning trove of documents.

I pressed forward with my reporting, calling the New York State Health Department to inquire about what had come of Tyler's anonymous complaint. It had been forwarded to the federal Centers for Medicare and Medicaid Services for investigation, I was told. But when I called CMS, I learned that no one there could find any trace of it. It had somehow been lost in the shuffle. To their credit, the folks who ran the agency's lab-oversight division seemed serious about following up on it now that they knew of its existence. They asked me to forward it to them and assured me it wouldn't be overlooked this time.

Meanwhile, Matthew Traub was continuing to give me the runaround. It seemed I was the only reporter in America to whom Holmes wouldn't grant an interview. She had recently appeared on CBS's morning news program, Fareed Zakaria's show on CNN, and Jim Cramer's **Mad Money** on CNBC. The icing on the cake was when one evening in early June I glanced up from my computer at one of the TVs in the newsroom and there she was in her black turtleneck carrying forth on **Charlie Rose**. During a heated phone conversation the next day, I told Traub that Theranos couldn't keep putting me off indefinitely. If not Holmes, someone from the company needed to meet with me to address my questions and it had to happen soon, I yelled, pacing back and forth in front of my stoop in Brooklyn.

Traub came back to me a few days later proposing that I meet with a Theranos representative at the offices of Boies Schiller in Manhattan. I initially agreed but then thought the better of it. That would be the equivalent of marching straight into the lion's den. I called him back and told him the Theranos representative—and the phalanx of attorneys I suspected would accompany him—needed to come to me. A meeting was scheduled for 1:00 p.m. on Tuesday, June 23, at 1211 Avenue of the Americas, home to the **Wall Street Journal**.

| TWENTY-ONE |

Trade Secrets

The Theranos delegation that came to the **Journal**'s offices was made up mostly of lawyers. Leading the pack was David Boies. He was flanked by Mike Brille, Meredith Dearborn, and Heather King, a former Boies Schiller partner and Hillary Clinton aide who had become Theranos's general counsel less than two months earlier. Rounding out the group were Matthew Traub and Peter Fritsch, a former **Journal** reporter who was the cofounder of an opposition research firm in Washington, D.C. The lone Theranos executive was Daniel Young.

Anticipating fireworks, I had brought along my editor, Mike Siconolfi, who headed the investigations team, and Jay Conti, the deputy general counsel of the **Journal**'s parent company who worked closely with the newsroom on sensitive journalistic matters. I had kept both apprised of my reporting and had let them in on the secret of who my sources were.

We sat down in a conference room on the fifth floor of the **Journal**'s newsroom. The tone was set from the start when King and Dearborn pulled out little tape recorders and placed them at each end of the conference table. The message was clear: they were approaching the meeting as a deposition in a future legal proceeding.

At Traub's request, I had sent a new set of eighty questions two weeks earlier that were to form the basis for our discussion. King opened the meeting by saying they were there to rebut the "false premises" embedded in those questions. Then she lobbed the first missile.

"It seems apparent to us that certainly one of your key sources is a young man named Tyler Shultz." She looked straight at me as she said it in what was clearly a rehearsed opening salvo to try to fluster me. I kept my poker face on and said nothing. They could suspect Tyler all they wanted, but I wasn't about to betray his confidence and give them the confirmation they were fishing for. She continued by denigrating Tyler as young and unqualified and then asserting that my other sources were disgruntled former employees who were equally unreliable. Mike interrupted her diatribe. We weren't going to disclose who our confidential sources were nor should Theranos presume to know their identities, he said politely but firmly.

Boies chimed in for the first time, playing the good cop to King's bad cop. "We really just want to go through this step-by-step so that you see that there just really isn't a story here," the seventy-four-year-old superlawyer said softly. With his bushy eyebrows and thinning gray hair, he called to mind a grandfather trying to reconcile squabbling children.

I suggested we begin addressing the questions I had sent, but, before I had time to read the first one, King's demeanor turned aggressive again and she issued a sharp warning: "We do not consent to your publication of our trade secrets."

We were minutes into the meeting and it had become clear to me that her main strategy was going to be to try to intimidate us, so I decided it was time to convey to her that that wasn't going to work.

"We do not consent to waiving our journalistic privileges," I snapped back.

My retort seemed to have the desired effect. She turned more conciliatory and we started going through my questions one by one with the understanding that Daniel Young, as the only Theranos official present, would be the one answering them. However, we were soon at loggerheads again.

After Young acknowledged that Theranos owned commercial blood analyzers, which he claimed the company used only for comparison purposes, rather than for delivering patient results, I asked if one of them was the Siemens ADVIA. He declined to comment, citing trade secrets. I then asked whether Theranos ran small finger-stick samples on the Siemens ADVIA with a special dilution protocol. He again invoked trade secrets to avoid answering the question but argued that diluting blood samples was common in the lab industry.

From there, the discussion went in circles. These questions were at the heart of my story, I said. If they weren't prepared to answer them, what was the point of our meeting? Boies replied that they were trying to be helpful but that they would not reveal what methods Theranos employed unless we signed nondisclosure agreements. Those were secrets Quest and LabCorp were desperately trying to find out by any means possible, including industrial espionage, he claimed.

As I continued to press for more substantive answers, Boies grew angry. He was suddenly no longer the amiable grandfatherly figure. He growled and flashed his teeth like an old grizzly bear. This was the David Boies who inspired fear in his courtroom adversaries, I thought to myself. He took a swipe at my reporting methods, saying I had asked some doctors loaded questions that were damaging to Theranos. That sparked a tense exchange between us. We stared each other down from across the table.

Jay Conti jumped in to defuse the situation but was soon sparring with King and Brille. Their argument took an almost farcical turn.

"It just feels like you want us to give you the formula for Coke in order to convince you that it doesn't contain arsenic," King said.

"Nobody's asked for the formula for Coke!" Jay replied, annoyed.

More arguing ensued over the issue of what legitimately constituted a trade secret. How could anything involving a commercial analyzer manufactured by a third party possibly be deemed a Theranos trade secret? I asked. Brille replied unconvincingly that the distinction wasn't as simple as I made it out to be.

We turned to my questions about the Edison. How many blood tests did Theranos perform on the device? That too was a trade secret, they said. I felt like I was watching a live performance of the Theater of the Absurd.

Did Theranos really have any new technology? I asked provocatively.

Boies's temper flared again. Testing tiny fingerstick samples was something no one in the laboratory industry had been able to do before, he fumed. "Theranos is doing it and, unless it's magic, it's a new technology!"

"It sounds like the **Wizard of Oz,**" Jay quipped.

We continued going around in circles, never getting a straight answer about how many tests Theranos performed on the Edison versus commercial analyzers. It was frustrating but also a sign that I was on the right track. They wouldn't be stonewalling if they had nothing to hide.

The meeting dragged on in this manner for four more hours. As we continued going through my list of questions, Young did answer some of them without invoking trade secrets. He acknowledged problems with Theranos's potassium test but claimed they had quickly been solved and none of the faulty results had been released to any patients. Alan Beam had told me otherwise, so I suspected Young was lying about that. Young also confirmed that Theranos conducted proficiency testing differently than most laboratories but argued this was justified by the uniqueness of its technology. He also confirmed that the CLIA inspector hadn't seen the Normandy part of Theranos's lab during her inspection but claimed she had been informed of its existence.

One of his answers struck me as odd. When I brought up the **Hematology Reports** study Holmes had coauthored, Young immediately dismissed it as outdated. It had been conducted with older Theranos technology and its data were ancient, dating back to 2008, he said. Why then had Holmes cited it to **The New Yorker**? I wondered. It seemed Theranos was now distancing itself from it, probably because it was conscious of its flimsiness.

I asked about Ian Gibbons. Young acknowledged that Ian had been an important contributor in the company's early years but said his behavior had become erratic at the end of his life and implied that he was no longer in the know at that stage. King interjected, dismissing Gibbons as an alcoholic. Boies, meanwhile, attacked Rochelle Gibbons's credibility by pointing out that she had failed to provide a sworn declaration in the Fuisz case, leading the judge to rule against allowing her testimony at trial.

Whether she'd provided a sworn statement in the Fuisz case or not was beside the point, I told him. I found her credible as a source and she was speaking to me on the record.

"She's under oath with me," I said.

We eventually turned to the instances of questionable test results I had gathered during my reporting. To be able to respond to my specific patient examples, King said Theranos would need to obtain signed releases from each waiving his or her patient-privacy rights. She asked me to help gather the waivers from the patients. I agreed to do so.

By the time the meeting finally broke up, it was nearly six p.m. and King looked like she wanted to plant a dagger in my chest.

THREE DAYS LATER, Erika Cheung was working late in the lab at her new employer, a biotech firm called Antibody Solutions, when a colleague came over to tell her that a man was asking to see her.

The man had been waiting in his car in the parking lot for a long time, the colleague said.

Erika was immediately on her guard. Mona Ramamurthy, the head of human resources at Theranos, had left several messages on her cell-phone voicemail earlier in the day saying there was something urgent she needed to discuss with her. Erika hadn't returned her calls and now some mysterious man was outside waiting to talk to her. She suspected the two were connected.

It was six p.m. on a Friday and not many people were left at Antibody Solutions' Sunnyvale offices. To be safe, Erika asked her colleague to walk her to her car. As they exited the building, a young man stepped out of an SUV and walked toward them at a fast clip with an envelope in his hand. He handed it to Erika, then turned around and left.

When she saw the address on the envelope, Erika's heart nearly stopped.

Via Hand Delivery
Ms. Erika Cheung
926 Mouton Circle
East Palo Alto, California 94303

The only person who knew that was where she was living was her colleague Julia. Two weeks earlier, Erika had let the lease expire on her apartment in Oakland and temporarily moved in with Julia ahead of a planned move to China in the fall. She'd been staying there only on weeknights, going camping or traveling on weekends. Not even her mother knew this address. The only way to have known it was to have followed her.

The letter inside the envelope was on Boies Schiller letterhead. As Erika read it, her sense of panic only increased:

> Dear Ms. Cheung,
>
> This firm represents Theranos, Inc. ("Theranos" or the "Company"). We have reason to believe that you have disclosed certain of the Company's trade secrets and other confidential information without authorization. We also have reason to believe that you have done so in connection with making false and defamatory statements about the Company for the purpose of harming its business. You are directed to immediately cease and desist from these activities. Unless this matter is resolved in accordance with the terms set forth in this letter by 5:00 p.m. (PDT) on Friday, July 3, 2015, Theranos will consider all appropriate remedies, including filing suit against you.

The letter went on to say that, if Erika wished to avoid litigation, she must submit to an interview with Boies Schiller attorneys and reveal what information she had disclosed about Theranos and to whom. It was signed by David Boies. Erika drove to Julia's house and stayed there all weekend with the blinds closed, too terrified to set foot outside.

BACK ON THE OTHER COAST, I was beginning to sense that things were escalating. That same Friday evening, I got a text from Alan Beam. It was the first I'd heard from him in nearly two months.

"Theranos is threatening me again," he wrote. "Their lawyers say they suspect I'm violating my affidavit."

We got on the phone and I filled him in on the marathon meeting with the Theranos delegation at the **Journal** a few days earlier. Rather than scaring him like I worried it might, Alan was fascinated by this new development. He had consulted with a new lawyer, a former federal prosecutor who had worked on the Medicare Fraud Strike Force, and felt less vulnerable to Theranos's intimidation tactics. In fact, he seemed to have changed his mind and to want to resume helping me get the story out.

Later that night, an email from Meredith Dearborn landed in my in-box. Attached to it was a formal letter from David Boies addressed to Jay Conti, who was the email's primary recipient. Citing several California statutes, the letter sternly demanded that the **Journal** "destroy or return" all Theranos trade secrets and confidential information in its possession. Even though Boies must have known there was zero chance we would do that, it was a shot across the bow.

Any remaining doubts I had that Theranos was waging an aggressive counterattack were put to rest the following Monday morning. I was sitting in my idled car listening to the radio while waiting for the street-sweeping truck to go by—one of the less pleasant aspects of life in Brooklyn—when my cell phone rang. I turned the volume down on the car radio and answered.

It was Erika and she seemed badly shaken. She told me about the man in the SUV, the address on the envelope, and the ultimatum from Boies. I tried to calm her down. Yes, it was highly likely she was under surveillance, I admitted. But I was sure that it had started only recently and that Theranos had no proof she was one of my sources. This was an attempt to smoke her out, I said. They were bluffing. I encouraged her to ignore the letter and to go about her business as usual. I could tell from her halting voice that she was still petrified, but she agreed to follow my advice.

The next day, I received an email from Dr. Sundene in Phoenix. A Theranos sales representative had come by her office to tell her that the company's president, Sunny Balwani, was in town and wanted to meet with her. When she'd declined the invitation, he had turned hostile and suggested that her refusal would have negative consequences. I couldn't believe it. Going after my confidential sources was one thing, but threatening a doctor who had spoken to me on the record was beyond the pale. I sent Heather King an email letting her know that I was aware of the sales representative's visit to Dr. Sundene's office and that, if I learned of any more such incidents, I would consider them newsworthy and would include mention of them in my story. King denied that the sales representative had done anything wrong.

Far from backing off, Theranos stepped it up a notch. Later that week, Boies sent the **Journal** a second letter. Unlike the first one, which was just two pages long, this one ran twenty-three pages and explicitly threatened a lawsuit if we published a story that defamed Theranos or disclosed any of its trade secrets. Much of the letter was a searing assault on my journalistic integrity. In the course of my reporting, I had "fallen far short of being fair, objective, or impartial" and instead appeared hell-bent on "producing a predetermined (and false) narrative," Boies wrote.

His main evidence to back up that argument was signed statements Theranos had obtained from two of the other doctors I had spoken to claiming I had mischaracterized what they had told me and hadn't made clear to them that I might use the information in a published article. The doctors were Lauren Beardsley and Saman Rezaie from the Scottsdale practice I'd visited.

The truth was that I hadn't planned on using the

patient case Drs. Beardsley and Rezaie had told me about because it was a secondhand account. The patient in question was being treated by another doctor in their practice who had declined to speak to me. But, while their signed statements in no way weakened my story, the likelihood that they had caved to the company's pressure worried me.

I noticed there was no signed statement from Adrienne Stewart, the third doctor I had interviewed at that practice. That was a good thing because I planned on using one or both of the patient cases she had discussed with me. When I reached her by phone, she said that she was visiting family in Indiana and hadn't been present when Theranos representatives came by the practice. I told her about her colleagues' signed statements and warned her that the company would probably try the same heavy-handed tactics with her when she returned.

Dr. Stewart emailed a few days later to let me know that Balwani and two other men had indeed come by to speak to her as soon as she'd gotten back to Arizona. The receptionist had told them she was busy with patients, but they had refused to leave and had stayed in the waiting room for hours until she finally came out to shake their hands. They had made her agree to meet with them the following Friday morning, which was in two days. I had a bad feeling about that meeting, but there was nothing I could do about it. Dr. Stewart promised she wouldn't bow to any pressure. She felt it was important to take a stand for her patients and the integrity of lab testing.

When Friday arrived, I tried to check in with Dr. Stewart several times in the morning but couldn't reach her. She called back in the early evening, as I was driving out to eastern Long Island for the weekend with my wife and three kids. She sounded rattled. She told me Balwani had tried to make her sign a statement similar to the one her colleagues had signed, but she had politely refused. Furious, he had threatened to drag her reputation through the mud if she appeared in any **Journal** article about Theranos. Her voice trembling, she pleaded with me to no longer use her name. As I tried to reassure her that it was an empty threat, it dawned on me that there was nothing these people would stop at to make my story go away.

| TWENTY-TWO |

La Mattanza

The early days of July 2015 brought two pieces of good news for Theranos. The first was that the FDA had approved the company's proprietary finger-stick test for HSV-1, one of two strains of herpes virus. The second was that a new law Arizona had passed allowing its citizens to get their blood tested without a doctor's order—a bill Theranos had practically written itself and heavily lobbied for—was about to go into effect.

The company celebrated these milestones by throwing a Fourth of July party at the new headquarters on Page Mill Road. The festivities started in the cafeteria with rousing speeches from Holmes and Balwani and then moved outside to the building's courtyard, where an open bar, catered food, and techno music awaited employees.

Theranos touted the herpes test approval as proof that its technology worked, but I remained deeply skeptical. In laboratory parlance, the herpes test was a qualitative test. Such tests provided simple yes-or-no answers to the question of whether a person had a certain disease. They were technically much easier to get right than quantitative tests designed to measure the precise amount of an analyte in the blood. Most routine blood tests were quantitative ones.

I called a source of mine who was high up in the FDA's medical-device division. He confirmed my thinking. The herpes test approval was a one-off clearance that was in no way a blanket endorsement of Theranos's technology, he said. In fact, the clinical data the company had submitted to the agency for a number of its other finger-stick tests were poor and wouldn't pass muster, he added. When in turn I told him about the things I had learned in the course of my reporting, ranging from Theranos's practice of running diluted finger-stick samples on commercial analyzers to its gaming of proficiency testing and the questionable test results some doctors and patients had received, he sounded disturbed.

Part of the problem was that, three years after Holmes's clash with the now-retired Lieutenant Colonel David Shoemaker, Theranos continued to operate in a regulatory no-man's-land. By using its proprietary devices only within the walls of its own laboratory and not seeking to commercialize them, it was able to continue to avoid close FDA scrutiny. At the same time, it gave the appearance of cooperating with the agency by publicly supporting its drive to regulate laboratory-developed tests and voluntarily submitting some of its own LDTs, like the herpes test, to it for approval.

My source said it was hard for the agency to take any adverse action against a company that portrayed itself as the lab world's biggest advocate of FDA regulation, especially one as politically connected as Theranos. At first, I thought he was referring to its board of directors, but that was the least of his concerns. He pointed out how chummy Holmes had gotten with the Obama administration. He had seen her at the launch of the president's precision medicine initiative earlier in the year, one of several White House appearances she'd made in recent months. The latest had been a state dinner in honor of Japan's prime minister, where she was photographed in a body-hugging black gown on the arm of her brother. Despite all this, his parting words made me think Theranos might not be able to fool the FDA much longer: "I'm very concerned about what they're doing."

OVER AT FORTUNE, Roger Parloff had a different take on the herpes test approval than I did. In an article he published on the magazine's website, he wrote that it was "a strong endorsement of the integrity of" Theranos's methods.

During a phone interview Holmes granted him for this second article, Parloff inquired about an Ebola test Theranos had in the works. George Shultz had made a passing mention of it at a conference a few months earlier. Given that an Ebola epidemic had been raging in West Africa for more than a year, Parloff thought a rapid finger-stick test to detect the deadly virus could be of great use to public health authorities and had been interested in writing about it. Holmes said she expected to obtain emergency-use authorization for the test shortly and invited him to come see a live demonstration of it at Boies Schiller's Manhattan offices.

A few days later, Parloff arrived at the law firm and was greeted by Dan Edlin, one of Christian Holmes's Duke fraternity brothers. Edlin showed him to a conference room where two black Theranos devices had been set up side by side (they were miniLabs, not Edisons). For reasons Parloff didn't understand, Holmes had wanted the demo to include a potassium test too (no doubt because I had been asking tough questions about that particular test). So Edlin drew blood from Parloff's finger twice. One machine would perform the

Ebola test and the other the potassium test, he explained. Parloff wondered fleetingly why one of the devices couldn't simultaneously perform both tests from a single blood sample but decided not to press the issue.

Parloff and Edlin made small talk while they waited for the test results. After about twenty-five minutes, the tests still hadn't been completed. Edlin said it was because the devices had just been installed and needed to warm up. The tests' progress was represented by the darkening edge of a circle on the devices' digital screens, like app downloads on an iPhone. Inside the circle, a percentage number told the user how much of the test had been completed. Based on how slowly the edge of one of the circles was filling up, it looked to Parloff like it might take several more hours. He couldn't wait around that long. He told Edlin he needed to head back to work.

After Parloff left, Kyle Logan, the young chemical engineer who had won an academic award at Stanford named after Channing Robertson, entered the conference room. He'd flown in with Edlin on the red-eye from San Francisco that morning and was there to provide technical support. Noticing that the miniLab running the potassium test was stuck at 70 percent completion, he took the cartridge out and rebooted the machine. He had a pretty good idea what had happened.

Balwani had tasked a Theranos software engineer named Michael Craig to write an application for the miniLab's software that masked test malfunctions. When something went wrong inside the machine, the app kicked in and prevented an error message from appearing on the digital display. Instead, the screen showed the test's progress slowing to a crawl.

This is exactly what had happened with Parloff's potassium test. Luckily, enough of the test had occurred before the malfunction that Kyle was able to retrieve a result from the machine. The breakdown had happened while the device was running the test again on the control part of the sample. Normally, it would have been preferable to have the initial result confirmed by the control, but Daniel Young told Kyle over the phone that it was OK to do without it in this case.

In the absence of real validation data, Holmes used these demos to convince board members, prospective investors, and journalists that the miniLab was a finished, working product. Michael Craig's app wasn't the only subterfuge used to maintain the illusion. During demos at headquarters, employees would make a show of placing the finger-stick sample of a visiting VIP in the miniLab, wait until the visitor had left the room, and then take the sample out and bring it to a lab associate, who would run it on one of the modified commercial analyzers.

As for Parloff, he had no idea he'd been duped. That evening, he got an email from Theranos with a password-protected attachment containing his results. When he opened the attachment, he was happy to see that he'd tested negative for Ebola and that his potassium value was within the normal range.

BACK IN CALIFORNIA, Holmes and Balwani were laying the groundwork for a bigger show-and-tell. Holmes had invited Vice President Joe Biden to come visit Theranos's Newark facility, which was now home to both Theranos's clinical laboratory and its miniLab manufacturing operations.

It was an audacious move given that, since Alan Beam's departure in December 2014, the lab had been operating without a real director. To keep this hidden, Balwani had recruited a dermatologist named Sunil Dhawan to replace Beam on the lab's CLIA license. Although Dhawan had no degree or board certification in pathology, he technically met state and federal requirements because he was a medical doctor and had overseen a little lab affiliated with his dermatology practice that analyzed skin samples. The reality, however, was that he was unqualified to run a full-fledged clinical lab. Not that it mattered. Balwani only intended him to be a figurehead. Some lab employees in Newark never saw Dhawan in the building.

Not only was the lab leaderless but its morale was at rock bottom. Two months earlier, Balwani had terrorized its members after a scathing critique of Theranos appeared on Glassdoor, the website where current and former employees reviewed companies anonymously. Titled "A pile of PR lies," it read in part:

> Super high turnover rate means you're never bored at work. Also good if you're an introvert because each shift is short-staffed. Especially if you're swing or graveyard. You essentially don't exist to the company.
>
> Why be bothered with lab coats and safety goggles? You don't need to use PPE at all. Who cares if you catch something like HIV or Syphilis? This company sure doesn't!
>
> Brown nosing, or having a brown nose, will get you far.
>
> How to make money at Theranos:
>
> 1. Lie to venture capitalists
> 2. Lie to doctors, patients, FDA, CDC, government. While also committing highly unethical and immoral (and possibly illegal) acts.

Negative Glassdoor reviews about the company weren't unusual. Balwani made sure they were balanced out by a steady flow of fake positive reviews he ordered members of the HR department to write. But this particular one had sent him into a rage. After getting Glassdoor to remove it, he'd launched a witch hunt in Newark, conducting interrogations of employees he suspected of having written it. He was so mean to one of them, a woman named Brooke Bivens, that he made her cry. He never found the culprit.

More recently, Balwani had fired Lina Castro, a well-liked and respected member of the microbiology team. Lina's sin had been to push the company to institute standard environmental health and safety protections in the lab. The morning after he fired her, Balwani had bragged to the remaining members of her team that he was worth billions and that he came to work every day because he wanted to. Everyone else should feel the same way, he said, implying that Castro had been too negative and not committed enough to the Theranos mission.

As had been the case in the old Facebook building in Palo Alto, the lab's operations in Newark were divided between Jurassic Park and Normandy. The new Jurassic Park occupied a huge room with neon lights and vinyl flooring. Lab associates' desks were clustered in one corner beneath a giant flat-screen monitor that displayed a constant stream of inspirational quotes and complimentary customer reviews. The commercial analyzers used to process regular venous samples dotted the rest of the space. Normandy occupied another room crammed with dozens of black-and-white Edisons and the Siemens machines Daniel Young and Sam Gong had hacked.

Holmes and Balwani wanted to impress the vice president with a vision of a cutting-edge, completely automated laboratory. So instead of showing him the actual lab, they created a fake one. They made the microbiology team vacate a third, smaller room, had it repainted, and lined its walls with rows of miniLabs stacked up on metal shelves. Since most of the miniLabs that had been built were in Palo Alto, they had to be transported back across the bay for the stunt. The members of the microbiology team weren't sure why they were being moved at first, but they figured it out when a Secret Service advance team showed up a few days before Biden arrived.

The day of the visit, most members of the lab were instructed to stay home while a few local news photographers and television cameras were allowed into the building to ensure the event got some press. Holmes took the vice president on a tour of the facility and showed him the fake automated lab. Afterward, she hosted a roundtable about preventive health care on the premises with a half dozen industry executives, including the president of Stanford Hospital.

During the roundtable discussion, Biden called what he had just seen "the laboratory of the future." He also praised Holmes for proactively cooperating with the FDA. "I know the FDA recently completed favorable reviews of your innovative device," he said. "The fact that you're voluntarily submitting all of your tests to the FDA demonstrates your confidence in what you're doing."

A FEW DAYS LATER, on July 28, I opened that morning's edition of the **Journal** and nearly spit out my coffee: as I was leafing through the paper's first section, I stumbled across an op-ed written by Elizabeth Holmes crowing about Theranos's herpes-test approval and calling for all lab tests to be reviewed by the FDA. She'd been denying me an interview for months, her lawyers had been stonewalling and threatening my sources, and here she was using my own newspaper's opinion pages to perpetuate the myth that she was regulators' best friend.

Because of the firewall between the **Journal**'s news and editorial sides, Paul Gigot and his staff had no idea I was working on a big investigative piece about the company. So I couldn't blame them for publishing whatever they saw fit. But I was annoyed. I suspected Holmes was trying to use the positive editorial-page coverage to make it more difficult for the paper to publish my investigation.

In the meantime, Alan Beam was coming under renewed pressure from Boies's henchmen. They were threatening to report him for violations of HIPAA, the federal health privacy law, on the grounds that some of the emails he had forwarded to his Gmail account before resigning contained patient information. His new lawyer had to fend them off from London, where he was on vacation with his wife. Balwani was also beginning to harass some of the patients I had talked to, insisting that they get on the phone with him and giving them the third degree when they did.

I had filed a draft of my story a week earlier and decided to walk over to my editor's office to see where he stood with his edit. Once he was done with it, the story would be sent to the paper's page-one editor, who would assign it to someone on his team for a second, closer edit. Then the standards editor and the lawyers would comb through it line by line. It was a slow process that often took weeks and sometimes months. I wanted to speed it up. The longer we took to publish, the more time we gave Theranos to turn my sources.

Mike Siconolfi was his usual cheerful self when I popped my head into his office. He motioned for me to sit down. I told him I felt we should move faster. There was no telling what Theranos and Boies would try next. I pointed out Holmes's op-ed and Biden's ballyhooed visit to Theranos's Newark facility a few days earlier.

Mike cautioned patience. This story was a bombshell and we needed to make sure it was bulletproof when we went to press with it, he said. Mike was of Italian American heritage and he loved using Italian metaphors. I had heard him tell the story of his ancestor Prince Siconulf, who ruled the region surrounding the Amalfi Coast in the ninth century, about ten times.

"Did I ever tell you about **la mattanza**?" he asked. Oh boy, here we go again, I thought.

He explained that **la mattanza** was an ancient Sicilian ritual in which fishermen waded into the Mediterranean Sea up to their waist with clubs and spears and then stood still for hours on end until the fish no longer noticed their presence. Eventually, when enough fish had gathered around them, someone gave an imperceptible signal and in a split second the scene went from preternatural quiet to gory bloodbath as the fishermen struck viciously at their unsuspecting quarry. What we were doing was the journalistic version of **la mattanza,** Mike said. We were patiently lying in wait until we were ready to publish and then, at some time of our choosing, we would strike. As he said this, he mimicked a Sicilian fisherman violently wielding his spear, which made me laugh.

I told him that I was on board with the **mattanza** approach as long as the story ran before Holmes's appearance at the **Journal**'s annual technology conference in Laguna Beach in October. I had recently gotten wind that she was on the conference's list of guest speakers and felt that it would put the paper in an impossible position if my article hadn't been published by then. Mike agreed. The conference was two and a half months away. That gave us ample time, he said.

| TWENTY-THREE |

Damage Control

Meanwhile, behind the scenes, Holmes was trying another avenue to quash the story.

In March, a month after I had started digging into the company, Theranos had closed another round of funding. Unbeknownst to me, the lead investor was Rupert Murdoch, the Australian-born media mogul who controlled the **Journal**'s parent company, News Corporation. Of the more than $430 million Theranos had raised in this last round, $125 million had come from Murdoch. That made him the company's biggest investor.

Murdoch had first met Holmes in the fall of 2014 at one of Silicon Valley's big galas, the annual Breakthrough Prize dinner. Held in Hangar 1 of NASA's Ames Research Center in Mountain View, the award honors outstanding contributors to the fields of the life sciences, fundamental physics, and mathematics. It was created by the Russian technology investor Yuri Milner with Facebook founder Mark Zuckerberg, Google cofounder Sergey Brin, and Chinese tech tycoon Jack Ma. During the dinner, Holmes came over to Murdoch's table, introduced herself, and chatted him up. The strong first impression she made on him was bolstered by Milner, who sang her praises when Murdoch later asked him what he thought of the young woman.

They met again a few weeks later at the media mogul's Northern California ranch. Murdoch, who had only one bodyguard, was surprised by the size of the security detail Holmes arrived with. When he asked her why she needed it, she replied that her board insisted on it. Over a lunch served by the ranch's staff, Holmes pitched Murdoch on an investment, emphasizing that she was looking for long-term investors. Don't expect any quarterly reports for a while, she warned him, and certainly not an initial public offering. The investment packet that was later delivered to Murdoch's Manhattan office reiterated that message. Its cover letter stated in the first paragraph that Theranos planned to remain private for the "long term" and went on to repeat those two words no fewer than fifteen times.

Murdoch was known to dabble in Silicon Valley startup investments. He'd been an early investor in Uber, turning a $150,000 bet into some $50 million.

But unlike the big venture capital firms, he did no due diligence to speak of. The eighty-four-year-old mogul tended to just follow his gut, an approach that had served him well on his way to building one of the world's biggest media and entertainment empires. The one call he placed before investing in Theranos was to Toby Cosgrove, the CEO of the Cleveland Clinic. Holmes had mentioned that she was on the verge of announcing an alliance with the world-renowned heart center. Like Yuri Milner, Cosgrove had only good things to say when Murdoch reached him.

Theranos was by far the single biggest investment Murdoch had ever made outside of the media assets he controlled, which included the 20th Century Fox movie studio, the Fox broadcast network, and Fox News. He was won over by Holmes's charisma and vision but also by the financial projections she gave him. The investment packet she sent forecast $330 million in profits on revenues of $1 billion in 2015 and $505 million in profits on revenues of $2 billion in 2016. Those numbers made what was now a $10 billion valuation seem cheap.

Murdoch also derived comfort from some of the other reputable investors he heard Theranos had lined up. They included Cox Enterprises, the Atlanta-based, family-owned conglomerate whose chairman, Jim Kennedy, he was friendly with, and the Waltons of Walmart fame. Other big-name investors he didn't know about ranged from Bob

Kraft, owner of the New England Patriots, to Mexican billionaire Carlos Slim and John Elkann, the Italian industrialist who controlled Fiat Chrysler Automobiles.

By the time Mike Siconolfi and I had our conversation about the ancient art of Sicilian fishing in late July, Holmes had had three private meetings with Murdoch. The latest had taken place earlier that month, when she'd hosted him in Palo Alto and showed him the miniLab. During the visit, she'd raised my story, telling him the information I had gathered was false and would do great damage to Theranos if it was published. Murdoch had demurred, saying he trusted the paper's editors to handle the matter fairly.

In late September, as we were getting close to publication, Holmes met with Murdoch a fourth time in his office on the eighth floor of the News Corporation building in Midtown Manhattan. My desk in the **Journal**'s newsroom was just three floors below, but I had no idea she was on the premises. She brought up my story with renewed urgency, hoping Murdoch would offer to kill it. Once again, despite the substantial investment he had at stake, he declined to intervene.

WHILE HOLMES TRIED unsuccessfully to sway the **Journal**'s owner, Theranos continued its scorched-earth campaign against my sources.

Boies Schiller's Mike Brille sent a letter to Rochelle Gibbons threatening to sue her if she didn't cease making what he termed "false and defamatory statements" about the company and its executives. In Phoenix, two new patients showed up for appointments at Dr. Sundene's office and threw tantrums. She had to hire an attorney to get Yelp to take down incendiary reviews they posted on the site about her. I had managed to keep Dr. Stewart from succumbing to Balwani's pressure, but Theranos had convinced her practice to accept in-office laboratory services from it to blunt her account of inaccurate test results.

Still, other on-the-record sources such as Dr. Gary Betz, the nurse Carmen Washington, and Maureen Glunz, the patient who had spent hours in the emergency room on the eve of Thanksgiving, remained impervious to the company's intimidation tactics. And Alan Beam and Erika Cheung continued to cooperate with the story as confidential sources, as did several other former employees.

While Tyler Shultz remained unreachable (I'd gotten his mother on the phone and left a message for him with her, to no avail), I assumed Theranos would have presented us with a signed statement similar to the ones from Drs. Rezaie and Beardsley if it had succeeded in making him recant. Besides, there was nothing it could do to make the emails he had given me disappear. Those spoke for themselves.

In a last-ditch effort to prevent publication, Boies sent the **Journal** a third lengthy letter, reiterating his threat to sue the paper and dismissing my reporting as an elaborate fantasy concocted by a fertile mind:

> I have tried to figure out how we could have arrived at a place where **The Journal** is considering publication of an article that we know to be false, misleading, and unfair, and that threatens to disclose information that Theranos rigorously protects as trade secrets.
>
> The root of the problem may be the drama of the reporter's original thesis, which may fall into the category of "too good to check." That thesis, as Mr. Carreyrou explained in discussions with us, is that all of the recognition by the academic, scientific, and health-care communities of the breakthrough contributions of Theranos' achievements is wrong; that every previous published report about Theranos, including in **The Journal** itself, has been the result of misleading manipulation by the company; and that the company and its founder are essentially perpetrating a fraud by touting a technology that does not work and using existing commercial equipment to do tests that Theranos pretends are done with new technology. Certainly such an exposé, if true, would be a powerful piece of investigative journalism. The

> problem may be that even though that thesis is not true, it is just too dramatic to let go.

The letter requested an audience with Gerry Baker, the **Journal**'s editor in chief. For the sake of fairness, Baker granted him one but made sure to invite me and Mike to attend as well as Jay Conti and Neal Lipschutz, the paper's standards editor.

At 4:00 p.m. on Thursday, October 8, we met with Boies again in a different conference room on the sixth floor of the **Journal**'s newsroom. This time, he came with a smaller contingent made up of Heather King and Meredith Dearborn. As she had in June, King pulled out a little tape recorder and set it on the table between us.

Although they continued to argue strenuously that my reporting was flawed and inaccurate, Boies and King made two key admissions during this second meeting that strengthened our hand. Acknowledging for the first time that Theranos didn't run all of its blood tests on its proprietary devices, Boies described the transition to doing so as "a journey" that would take the company some time to complete. The second came after I brought up several recent wording changes I'd noticed on the Theranos website. One in particular seemed telling: the sentence "Many of our tests require only a few drops of blood" had been deleted. When I asked why, King inadvertently blurted out that she assumed it was for "marketing accuracy."

(Later, she would insist she never pronounced those words.)

Toward the end of the meeting, Boies tried one last tack: if we were willing to delay going to press a little longer, he would arrange a demonstration of the Theranos device. They had done one for **Fortune** magazine not long before, he said, so there was no reason they couldn't do one for us too. Such a demonstration would provide the incontrovertible proof that we were wrong about the machine not working, Boies contended.

Mike and I asked how soon it could take place, which tests would be administered, and what assurance we would have that the results would come from the device and not involve some sleight of hand. When Boies replied that it would probably take several weeks to organize and was wishy-washy on the other points, Baker politely declined the offer. He shared our view that we had to publish before Holmes's appearance at the **Journal**'s tech conference, which was less than two weeks away.

Baker told Boies that we weren't going to wait weeks but that he was willing to push back publication by a few more days to give Holmes one last opportunity to talk to me. He gave her until early the following week to pick up the phone and call me. She never did.

THE STORY WAS PUBLISHED on the **Journal**'s front page on Thursday, October 15, 2015. The headline, "A Prized Startup's Struggles," was understated but the article itself was devastating. In addition to revealing that Theranos ran all but a small fraction of its tests on conventional machines and laying bare its proficiency-testing shenanigans and its dilution of finger-stick samples, it raised serious questions about the accuracy of its own devices. It ended with a quote from Maureen Glunz saying that "trial and error on people" was "not OK," bringing home what I felt was the most important point: the medical danger to which the company had exposed patients.

The story sparked a firestorm. NPR interviewed me on its **Marketplace** program first thing in the morning. The editor of **Fortune,** the publication that had done more than any other to elevate Holmes to stardom, made the story the focus of his daily email to readers. "A high-flying unicorn has been brought closer to earth this morning by a deeply reported story on the front page of the **Wall Street Journal,**" he wrote. **Forbes** and **The New Yorker,** two other magazines that had played roles in Holmes's rise to fame, also picked up the story, as did many other news outlets.

In Silicon Valley, it became the talk of the town. Some venture capitalists reflexively jumped to Holmes's defense. One of them was former Netscape cofounder Marc Andreessen, whose wife had just profiled Holmes in a cover story for the **New York Times**'s style magazine headlined "Five Visionary Tech Entrepreneurs Who Are Changing the World." But others were less charitable, having long harbored doubts of their own. Why had Holmes always been so secretive about her technology? Why had she never recruited a board member with even basic knowledge of blood science? And why hadn't a single venture capital firm with expertise in health care put money into the company? For these observers, the story confirmed what they'd quietly suspected.

There was a third group of people who didn't know what to believe given Theranos's strong denials. In a press release it posted on its website, the company called the story "factually and scientifically erroneous and grounded in baseless assertions by inexperienced and disgruntled former employees and industry incumbents." It also let it be known that Holmes would appear on Jim Cramer's **Mad Money** program that evening to rebut the allegations.

We knew that the battle was far from over and that Theranos and Boies would be coming at us hard in the following days and weeks. Whether my reporting stood up to their attacks would largely depend on what actions, if any, regulators took. Rumors had been circulating among former Theranos employees of an FDA inspection, but I hadn't been able to confirm it by the time we went to press. I had called my source at the agency several times but hadn't been able to reach him.

I decided to try him again that day before lunch. This time he picked up the phone. On deep background, he confirmed to me that the FDA had recently conducted a surprise inspection of Theranos's facilities in Newark and Palo Alto. Dealing a severe blow to the company, the agency had declared its nanotainer an uncleared medical device and forbidden it from continuing to use it, he said.

He explained that the FDA had targeted the little tube because, as a medical device, it clearly fell under its jurisdiction and gave it the most solid legal cover to take action against the company. But the underlying reason for the inspection had been the poor clinical data Theranos had submitted to the agency in an effort to get it to approve its tests. When the inspectors had failed to find any better data on-site, the decision had been made to shut down the company's finger-stick testing by taking away the nanotainer, he said. That wasn't all: he said the Centers for Medicare and Medicaid Services had also just launched its own inspection of Theranos. He didn't know whether it was still ongoing but was sure it spelled more trouble for the company. Mike and I discussed these revelations and quickly went to work on a follow-up story for the next day's paper.

A few hours later, I was standing over the shoulder of the page-one editor who was handling my new story when Holmes's face appeared on a nearby TV tuned to CNBC. We took a break from the edit and turned up the volume. Dressed in her usual all-black attire and sporting a strained smile, she played the role of the visionary Silicon Valley innovator who was being smeared by entrenched interests trying to thwart progress. "This is what happens when you work to change things," she said. "First they think you're crazy, then they fight you, and then all of a sudden you change the world." But, when Jim Cramer asked her about specific elements of the article, such as the company's use of third-party analyzers for most of its tests, she turned defensive and gave evasive and misleading answers.

I had sent Heather King an email earlier in the day to let her know I was working on a second story and to request a Theranos comment about the things I was going to report. King hadn't replied. I now knew why: toward the end of her interview with Cramer, Holmes dropped mention of the nanotainer withdrawal and spun it as a voluntary decision. She was trying to get ahead of my scoop.

We quickly published my follow-up piece online. Setting the record straight, it revealed that the FDA had forced the company to stop testing blood drawn from patients' fingers and declared its nanotainer an "unapproved medical device." The story made the front page of the paper's print edition the next morning, providing more fuel to what was now a full-blown scandal.

HOLMES WASN'T IN Palo Alto the day our first story was published. She was attending a meeting of Harvard Medical School's board of fellows. She did her CNBC interview that evening from Boston. It wasn't until the next day that she flew back to California to address the growing crisis.

Theranos had issued a second press release that morning that amounted to what we in the news business call a "nondenial denial." "We are disappointed to see that **The Wall Street Journal** still can't get its facts straight," it began, before going on to admit that the company had "temporarily" withdrawn its little blood tubes in what it portrayed as a proactive move to seek FDA clearance for their use.

In the late afternoon, an email went out to all company employees instructing them to gather in the cafeteria of the Page Mill Road building for a meeting. Holmes wasn't her usual well-put-together self. Her hair was disheveled from her travels and she wore glasses instead of contact lenses. Standing next to her were Balwani and Heather King. Striking a defiant tone, she told the assembled staff that the two articles the **Journal** had published were filled with falsehoods seeded by disgruntled former employees and competitors. This sort of thing was bound to happen when you were working to disrupt a huge industry with powerful incumbents who wanted to see you fail, she said. Calling the

Journal a "tabloid," she vowed to take the fight to the paper.

When she opened the floor to questions, Patrick O'Neill, the former advertising industry executive who had helped craft her trailblazing image, was one of the first to raise his hand.

"Do we really want to take on the **Wall Street Journal**?" he asked, incredulous.

"Not the **Journal,** the journalist," Holmes replied.

After she answered a few more questions, one of the senior hardware engineers asked Balwani if he would lead them in a chant. Everyone instantly knew what chant the engineer had in mind. Three months earlier, when the company had received its herpes test approval from the FDA, Balwani had exhorted employees to yell "Fuck you" in unison during a similar meeting in the cafeteria. At the time, the shouts had been directed at Quest and LabCorp.

Balwani was more than happy to indulge the engineer's request for an encore.

"We have a message for Carreyrou," he said.

At his signal, he and many of the several hundred employees in attendance chanted: "Fuck you, Carrey-rou! Fuck you, Carrey-rou!"

WHEN HOLMES HAD SAID she planned to take the fight to the **Journal,** she meant it.

Many people had assumed she would back out of the paper's WSJ D.Live conference the following week. But on the appointed day and at the appointed time, she appeared at the beachfront Montage hotel and resort in Laguna Beach with her platoon of guards and joined Jonathan Krim, the **Journal**'s technology editor, onstage. The audience of more than one hundred—a mixture of venture capitalists, startup founders, bankers, and public-relations executives who had paid five thousand dollars each to attend the three-day conference—buzzed in anticipation.

Mike Siconolfi had wanted me to handle the interview, but the paper didn't like the idea of making last-minute changes to an event that had taken months of planning. Besides, I couldn't leave New York. My wife was stuck on jury duty in a federal trial in Islip, Long Island, a two-hour drive from Brooklyn. I had to take care of our kids.

There was so much interest in the unfolding Theranos story that the **Journal** had decided to stream the interview live on its website. Several of us watched it in Neal Lipschutz's office.

Holmes came out swinging almost from the start. That was no surprise: we had expected her to be combative. What we hadn't fully anticipated was her willingness to tell bald-face lies in a public forum. Not just once, but again and again during the half-hour interview. In addition to continuing to insist that the nanotainer withdrawal had been voluntary, she said the Edison devices referred to in my stories were an old technology that Theranos hadn't used in years. She also denied that the company had ever used commercial lab equipment for finger-stick tests. And she claimed that the way Theranos conducted proficiency testing was not only perfectly legal, it had the express blessing of regulators.

The biggest lie, to my mind, was her categorical denial that Theranos diluted finger-stick samples before running them on commercial machines. "What the **Journal** described—that we take a sample, dilute it, and put it on a commercial analyzer—is inaccurate, and that's not what we do," she told Krim. "In fact, I bet you if you tried that, it wouldn't work because it's just not possible to dilute a sample and put it onto a commercial analyzer. I mean, there are so many things that are wrong with that." As I shook my head in disgust, a text flashed on my phone screen. It was from Alan Beam: "I can't believe what she just said!" he wrote.

From there, Holmes turned her sights on the former employees who had spoken to me, calling them "confused" and seizing on their anonymity to discredit them. She claimed that one of them had worked at Theranos for only two months back in 2005, which was a complete fabrication. All our confidential sources had worked at the company in recent times. In response to a question about Rochelle Gibbons, she reprised the line she'd used with her employees five days earlier, likening the **Journal** to "a tabloid magazine." And she referred to me as "some guy" who had reported "false stuff about us."

One problem she faced was that we were no longer the only ones raising questions about Theranos. Several prominent Silicon Valley figures had begun criticizing the company publicly. One of them was a well-known former Apple executive named Jean-Louis Gassée. A few days earlier, Gassée had published an item on his blog describing sharply discordant blood-test results he had received from Theranos and Stanford Hospital over the summer. Gassée had written Holmes to inquire about the discrepancies but had never received a response. When Krim raised Gassée's case, Holmes claimed to have never received his email. Now that it knew about his complaint, Theranos would reach out to him to try to understand what had happened, she said.

As for the other instances of inaccurate test results described in our first story, she dismissed them as a few isolated cases from which general conclusions could not and should not be drawn.

Soon after the interview ended, Theranos posted a long document on its website that purported to rebut my reporting point by point. Mike and I went over it with the standards editors and the lawyers and concluded that it contained nothing that undermined what we had published. It was another smokescreen. The paper put out a statement to say that it stood by my stories.

AFTER HOLMES'S APPEARANCE at the **Journal** conference, Theranos announced that it was making changes to its board of directors, which had been getting lampooned since the publication of my first story. George Shultz, Henry Kissinger, Sam Nunn, and the other aging ex-statesmen all left to join a new ceremonial body called a board of counselors. In their place, Theranos made a new director appointment that signaled an escalation of hostilities: David Boies.

Sure enough, within days, the **Journal** received letters from Heather King demanding that it retract the central elements of my first two articles, calling them "libelous assertions." A third letter followed demanding that the paper retain all documents in its possession concerning Theranos, "including emails, instant messages, drafts, informal files, handwritten notes, faxes, memoranda, calendar entries, voice mail and any other Records stored in hard copy, or any electronic form (including personal cell phones) or any other medium."

In an interview with **Wired,** Boies suggested that a defamation suit was likely. "I think enough has now been put on the record so people are chargeable with being knowledgeable with what the facts are," he told the magazine. Taking King and Boies at their words, the **Journal**'s legal department dispatched a technician to copy the contents of my laptop and phone in preparation for litigation.

But if Theranos thought this saber rattling would make us stand down, it was mistaken. Over the next three weeks, we published four more articles. They revealed that Walgreens had halted a planned nationwide expansion of Theranos wellness centers, that Theranos had tried to sell more shares at a higher valuation days before my first story was published, that its lab was operating without a real director, and that Safeway had walked away from their previously undisclosed partnership over concerns about its testing. With each new story came a new retraction demand from Heather King.

In a windowless war room set up on the second floor of the Page Mill Road building in Palo Alto, Holmes and her communications consultants discussed strategies for how to hit back against my reporting. One approach she favored was to portray me as a misogynist. To generate further sympathy, she suggested she reveal publicly that she had been sexually assaulted as a student at Stanford. Her advisers counseled against going that route, but she didn't abandon it entirely. In an interview with **Bloomberg Businessweek,** she suggested she was the victim of sexism.

"Until what happened in the last four weeks, I didn't understand what it means to be a woman in this space," she told the magazine. "Every article starting with, 'A young woman.' Right? Someone came up to me the other day, and they were like, 'I have never read an article about Mark Zuckerberg that starts with 'A young man.' "

In the same story, her old Stanford professor, Channing Robertson, dismissed questions about the accuracy of Theranos's testing as absurd, saying the company would have to be "certifiable" to go to market with a product that people's lives depended on knowing that it was unreliable. He also maintained that Holmes was a once-in-a-generation genius, comparing her to Newton, Einstein, Mozart, and Leonardo da Vinci.

Holmes, too, continued to embrace an exalted image of herself. In her acceptance speech at **Glamour** magazine's Women of the Year Awards at Carnegie Hall, she held herself up as a role model for young women. "Do everything you can to be the best in science and math and engineering," she urged them. "It's that that our little girls will see when they start to think about who do they want to be when they grow up."

There was only one way the charade would end and that was if CMS, the chief regulator of clinical laboratories, took strong action against the company. I needed to find out what had come of that second regulatory inspection.

| TWENTY-FOUR |

The Empress Has No Clothes

On a Saturday evening in late September, about three weeks before the **Journal** published my first story, an email arrived in the in-box of Gary Yamamoto, the veteran CMS field inspector who had dropped in unannounced at the old Facebook building in 2012 and lectured Sunny Balwani about lab regulations. Under the subject line "CMS Complaint: Theranos Inc.," it began:

> Dear Gary,
>
> I've been nervous to send or even write this letter. Theranos takes confidentiality and secrecy to an extreme level that has always made me scared to say anything . . . I'm ashamed in myself for not filing this complaint sooner.

The email was from Erika Cheung and it contained a series of allegations, ranging from scientific misconduct to sloppy lab practices. It also said that Theranos's proprietary devices were unreliable, that the company cheated on proficiency testing, and that it had misled the state inspector who surveyed its lab in late 2013. Erika closed the email by saying that she'd resigned from the company because she couldn't live with herself knowing that she could "potentially devastate someones [**sic**] life by giving them a false and deceiving result."

Yamamoto and his superiors at CMS took the complaint so seriously that the agency launched a surprise inspection of Theranos's laboratory less than three days later. On the morning of Tuesday, September 22, Yamamoto and another field inspector from CMS's San Francisco regional office named Sarah Bennett arrived at the Newark facility and explained that they were there to survey the lab. Men in dark suits wearing earpieces denied them entry and told them to wait in a small reception room.

After a while, Sunny Balwani, Daniel Young, Heather King, and Boies Schiller's Meredith Dearborn arrived. They took the two CMS inspectors to a conference room and insisted on giving them a PowerPoint presentation. Although it felt like a diversion tactic, Yamamoto and Bennett politely sat through it. As soon as it was over, they asked for a tour of the lab.

As they headed out of the conference room, they

were escorted by more men in dark suits with fingers pressed to their ears. King and Dearborn followed close behind, holding laptops and taking notes. When they got to the lab rooms, they noticed that their doors were equipped with fingerprint scanners and that there was a buzzing sound when you entered. It reminded Yamamoto of the door buzzers in liquor stores.

Yamamoto and Bennett had initially set aside two days for the inspection, but they found so many problems and Theranos was missing so much basic lab documentation that they concluded they would have to return. Balwani asked for a two-month reprieve. He claimed that the company's new fiscal year was about to start and that it was in the midst of raising new funding. They agreed to come back in mid-November.

When they returned, the **Journal** investigation had been published, ratcheting up pressure on the agency to take action. Yamamoto noticed that security was a bit lighter and that Holmes was there to greet them. Balwani and King were also present again, along with a different set of outside attorneys and some lab consultants. The inspectors split up: Yamamoto roamed through the lab rooms and peppered the lab's personnel with questions, followed everywhere he went by Balwani, while Bennett set up shop in a conference room where King and the other lawyers kept close tabs on her.

This time, they stayed four days. At one point,

Bennett asked to conduct a confidential interview with one of the lab associates who worked in Normandy and had direct experience with the Edisons. She was made to wait in a windowless room for a long time until a young woman finally appeared. As soon as she sat down, the woman asked for an attorney. She looked coached and afraid.

ERIKA CHEUNG AND I had remained in sporadic contact after her parking lot scare in late June, but I didn't know she'd gathered the courage to reach out to a federal regulator. When I first heard about the CMS inspection, I had no idea it had been triggered by her.

Throughout the fall of 2015 and into the winter of 2016, I tried to find out what the inspection had uncovered. After Yamamoto and Bennett completed their second visit in November, there were rumblings from former Theranos employees in contact with current ones that it hadn't gone well, but details were hard to come by. In late January, we were finally able to publish a story reporting that the CMS inspectors had found "serious" deficiencies at the Newark lab, citing sources familiar with the matter. How serious became clear a few days later when the agency released a letter it had sent the company saying they posed "immediate jeopardy to patient health and safety." The letter gave the company ten days to come up with a credible correction plan and warned that failing to come back into compliance quickly could cause the lab to lose its federal certification.

This was major. The overseer of clinical laboratories in the United States had not only confirmed that there were significant problems with Theranos's blood tests, it had deemed the problems grave enough to put patients in immediate danger. Suddenly, Heather King's written retraction demands, which had been arriving like clockwork after each story we published, stopped.

However, Theranos continued to minimize the seriousness of the situation. In a statement, it claimed to have already addressed many of the deficiencies and that the inspection findings didn't reflect the current state of the Newark lab. It also claimed that the problems were confined to the way the lab was run and had no bearing on the soundness of its proprietary technology. It was impossible to disprove these claims without access to the inspection report. CMS usually made such documents public a few weeks after sending them to the offending laboratory, but Theranos was invoking trade secrets to demand that it be kept confidential. Getting my hands on that report became essential.

I called a longtime source of mine in the federal government who had access to it. The most he was willing to do was read some passages over the phone. That was enough for us to report one of the inspection's most serious findings: the lab had continued to run a blood-clotting test for months despite repeated quality-control failures indicating that it was faulty. "Prothrombin time," as the test was known, was a dangerous test to get wrong because doctors relied on it to determine the dosage of blood-thinning medication they prescribed to patients at risk of strokes. Prescribing too much blood thinner could cause patients to bleed out, while prescribing too little could expose them to fatal clots. Theranos couldn't refute our story, but it argued once more that its proprietary technology was not at issue. The prothrombin time test had been performed on regular venous samples using commercial equipment, it said. When its back was against the wall, the company was willing to admit to using conventional analyzers if doing so could help maintain the illusion that its own devices worked.

To try to force CMS to release the inspection report, I filed a Freedom of Information Act request for any and all documents connected to the Newark lab inspection and requested that it be expedited. But Heather King continued to urge the agency not to make the report public without extensive redactions, claiming that doing so would expose valuable trade secrets. It was the first time the owner of a laboratory under the threat of sanctions had demanded redactions to an inspection report, and CMS seemed unsure how to proceed. With each passing day, I became concerned that the full inspection findings would never be released.

As the tug-of-war with Heather King over the inspection report dragged on, news surfaced that Holmes would be hosting a fund-raiser for Hillary Clinton's presidential campaign at Theranos's headquarters in Palo Alto. She had long cultivated a relationship with the Clintons, appearing at several Clinton Foundation events and forging a friendship with their daughter. The fund-raiser was later relocated to the home of a tech entrepreneur in San Francisco, but a photo from the event showed Holmes holding a microphone and speaking to the assembled guests with Chelsea Clinton at her side. With the election eight months away and Clinton considered the front-runner, it was a reminder of how politically connected Holmes was. Enough to make her regulatory problems go away? Anything seemed possible.

I went back to my source and, this time, cajoled him into leaking the whole inspection report to me. Running 121 pages long, the document was as damning as one could expect. For one thing, it proved that Holmes had lied at the **Journal**'s tech conference the previous fall: the proprietary devices Theranos had used in the lab were indeed called "Edison," and the report showed it had used them for only twelve of the 250 tests on its menu. Every other test had been run on commercial analyzers.

More important, the inspection report showed, citing the lab's own data, that the Edisons produced wildly erratic results. During one month, they had failed quality-control checks nearly a third of the time. One of the blood tests run on the Edisons, a test to measure a hormone that affects testosterone levels, had failed quality control an astounding 87 percent of the time. Another test, to help detect prostate cancer, had failed 22 percent of its quality-control checks. In comparison runs using the same blood samples, the Edisons had produced results that differed from those of conventional machines by as much as 146 percent. And just as Tyler Shultz had contended, the devices couldn't reproduce their own results. An Edison test to measure vitamin B_{12} had a coefficient of variation that ranged from 34 to 48 percent, far exceeding the 2 or 3 percent common for the test at most labs.

As for the lab itself, it was a mess: the company had allowed unqualified personnel to handle patient samples, it had stored blood at the wrong temperatures, it had let reagents expire, and it had failed to inform patients of flawed test results, among many other lapses.

Heather King tried to prevent us from publishing the report, but it was too late. We posted it on the **Journal**'s website and the accompanying story quoted a laboratory expert who said its findings suggested the Edisons' results were no better than guesswork.

The coup de grâce came a few days later when we obtained a new letter CMS had sent to Theranos. It said the company had failed to correct forty-three of the forty-five deficiencies the inspectors had cited it for and threatened to ban Holmes from the blood-testing business for two years. As with the inspection report, Theranos was desperately trying to keep the letter from becoming public, but a new source had contacted me out of the blue and leaked it to me.

When we reported news of the threatened ban, it was no longer possible for Holmes to downplay the gravity of the situation. She had to come out and say something, so she gave an interview to Maria Shriver on NBC's **Today** show in which she professed to be "devastated." But not enough, it seemed, to apologize to the patients she had put in harm's way. Watching her, I got the distinct impression that her display of contrition was an act. I still didn't sense any real remorse or empathy.

After all, Theranos's employees, its investors, and its retail partner, Walgreens, had all learned of the inspection's findings and the threatened ban by reading the **Wall Street Journal**. If Holmes was sincere about making things right, why had she tried so hard to suppress their disclosure?

IN MAY 2016, I returned to the San Francisco Bay Area to try to find out what had happened to Tyler Shultz. It was almost exactly a year to the day since we'd met at the beer garden in Mountain View. Erika had told me Tyler was working on a research project with a nanotechnology professor at Stanford, so I drove my rental car to Palo Alto and searched for him in Stanford's School of Engineering. After asking around, I finally located him in a room in the materials science building.

Tyler wasn't surprised to see me. Erika had given me his real email address and I'd written to him to let him know I was coming back through town. He'd been noncommittal about meeting with me. Now that I was there, though, he relented. We walked over to a nearby cafeteria to grab some lunch and slipped into easy banter.

Tyler seemed in good spirits. He told me he was part of a small group of researchers at Stanford that had teamed up with a Canadian company to compete in the multimillion-dollar Qualcomm Tricorder XPRIZE competition. They were trying to build a portable device capable of diagnosing a dozen diseases from a person's blood, saliva, and vital signs.

When our conversation turned to Theranos, his brows furrowed and he tensed up. He didn't want to discuss the subject in an open place within earshot of other people, he said. He suggested we walk back to the materials science building. We found an empty classroom there and sat down. His relaxed demeanor in the cafeteria had given way to palpable anxiety.

"My lawyers forbade me from talking to you, but I can't keep this bottled up anymore," he said.

I agreed to keep whatever he was about to tell me off the record and only to write about it in the future if he gave me his permission to do so.

For the next forty-five minutes, I listened dismayed as Tyler told me about the ambush at his grandfather's house and the months of legal threats he'd endured. Despite it all, he had never buckled. He had steadfastly refused to sign any document Boies Schiller had put to him. If not for his courage and the more than $400,000 his parents had spent on his attorneys, I might never have been able to get my first article published, I realized. I felt pangs of guilt for having put him through such an ordeal.

The most heartbreaking part of it all was Tyler's estrangement from his grandfather. George Shultz had continued to side with Holmes in spite of everything my reporting had revealed. He and Tyler hadn't seen each other for the better part of a year and they communicated only through lawyers. The previous December, the Shultzes had hosted a party at a penthouse apartment they owned in San Francisco to celebrate George's ninety-fifth birthday. Holmes had attended, but not Tyler.

Tyler had heard through his parents that his grandfather continued to believe in the promise of Theranos. In a complete about-face after years of intense secrecy, Holmes was going to unveil the inner workings of her technology at the American Association for Clinical Chemistry's annual meeting on August 1, 2016. George believed that her presentation would silence the doubters. Tyler didn't understand why he couldn't see through her lies. What would it take for him to finally accept the truth?

As we parted, Tyler thanked me for doggedly pursuing the story. He pointed out that Theranos had consumed the past four years of his life dating back to his summer internship at the company between his junior and senior years in college. I thanked him in turn for helping me get the story out and for withstanding the immense pressure under which he had been placed.

Not long afterward, Theranos contacted Tyler's lawyers and told them it knew about our meeting. Since neither of us had told a soul about it, we deduced that Holmes was having one or both of us followed. Fortunately, Tyler didn't seem too worried about it. "Next time maybe I'll take a selfie with you and send it her way to save her the trouble of hiring PIs," he quipped in an email.

I now suspected Theranos had had both of us under continuous surveillance for a year. And, more than likely, Erika Cheung and Alan Beam too.

HOLMES HAD TOLD Maria Shriver on the **Today** show that she took responsibility for the Newark lab's failings, but it was Balwani who suffered the consequences. Rather than take the fall herself, she sacrificed her boyfriend. She broke up with him and fired him. In a press release, Theranos dressed up his departure as a voluntary retirement.

A week later, we reported that Theranos had voided tens of thousands of blood-test results, including two years' worth of Edison tests, in an effort to come back into compliance and avoid the CMS ban. In other words, it had effectively admitted to the agency that not a single one of the blood tests run on its proprietary devices could be relied upon. Once again, Holmes had hoped to keep the voided tests secret, but I found out about them from my new source, the one who had leaked to me CMS's letter threatening to ban Holmes from the lab industry. In Chicago, executives at Walgreens were astonished to learn of the scale of the test voidings. The pharmacy chain had been trying to get answers from Theranos about the impact on its customers for months. On June 12, 2016, it terminated the companies' partnership and shut down all the wellness centers located in its stores.

In another crippling blow, CMS followed through on its threat to ban Holmes and her company from the lab business in early July. More ominously, Theranos was now the subject of a criminal investigation by the U.S. Attorney's Office in San Francisco and of a parallel civil probe by the Securities and Exchange Commission. In spite of all these setbacks, Holmes felt she still had one card to play to turn public opinion around: wow the world with a display of her technology.

ON A MUGGY SUMMER DAY at the beginning of August, more than 2,500 people crowded into the grand ballroom of the Pennsylvania Convention Center in Philadelphia. Most were laboratory scientists who had come to hear Holmes speak at the annual meeting of the AACC. "Sympathy for the Devil" by the Rolling Stones was playing on the public announcement system, a choice of music that didn't seem like a coincidence.

The invitation the association had extended to Holmes was highly controversial among its members. Some had argued forcefully that it should be withdrawn given the events of recent months. But the association's leadership had seen a chance to generate publicity and buzz for the normally staid scientific conference. It was proven right in that regard: several dozen journalists had made the trek to Philadelphia to watch the spectacle.

After a few introductory remarks from AACC president Patricia Jones, Holmes stepped up to the lectern. She wore a white blouse under a dark jacket. Gone was the black turtleneck, which had become an object of ridicule since the previous fall.

What followed wasn't so much a scientific presentation as it was a new product exhibit. Over the next hour, Holmes proceeded to unveil the machine that had been but a malfunctioning prototype when Theranos had gone live with its blood tests nearly three years earlier: the miniLab. Theranos's engineers and chemists had improved the device since that early model, but the company still hadn't conducted a full clinical study to prove that it worked reliably across a wide range of assays using blood pricked from fingertips. While Holmes's presentation included some data, much of it involved venous blood drawn from the arm. The little finger-stick data it contained covered only eleven blood tests and wasn't independently verified or peer-reviewed. CMS had just banned Holmes from running clinical laboratories, but no matter: she explained that the miniLab connected wirelessly to servers at Theranos headquarters and could be deployed directly in patients' homes, doctors' offices, or hospitals, doing away with thc need for a central lab.

In effect, she was pirouetting back to her original vision of portable blood-testing machines operated remotely through Wi-Fi or cellular networks. Of course, after everything that had happened, commercializing such a system without the FDA's approval was out of the question. And putting together the thorough studies the agency would want to see would take years. That was why she had tried to bypass the FDA in the first place.

The odds that Holmes could pull off this latest Houdini act while under criminal investigation were very long, but watching her confidently walk the audience through her sleek slide show helped crystallize for me how she'd gotten this far: she was an amazing saleswoman. She never once stumbled or lost her train of thought. She wielded both engineering and laboratory lingo effortlessly and she showed seemingly heartfelt emotion when she spoke of sparing babies in the NICU from blood transfusions. Like her idol Steve Jobs, she emitted a reality distortion field that forced people to momentarily suspend disbelief.

The spell was broken, however, during the question-and-answer session when Stephen Master, an associate professor of pathology at Weill Cornell Medical Center in New York and one of three panelists invited onstage to ask Holmes questions, pointed out that the miniLab's capabilities fell far short of the original claims she had made. His comment drew a burst of applause from the audience. Reverting to the chastened persona from her **Today** show interview, Holmes acknowledged that Theranos had a lot of work to do, as she put it, to "engage" with the laboratory community. But she stopped short again of apologizing or admitting fault.

When Dennis Lo, a pathology professor at the Chinese University of Hong Kong, later asked her how the miniLab differed from the technology the company had used in its laboratory on patient samples, she dodged the question. It was a giant issue to sidestep, yet the hundreds of assembled pathologists remained civil and respectful despite her evasiveness. There were no boos or catcalls. The decorum broke down only briefly when Holmes turned to leave the stage at the end of the Q&A. "You hurt people," a voice yelled out from the dispersing crowd.

IF HOLMES HAD HOPED to rehabilitate her image and change the media narrative by unveiling the miniLab, that hope was dashed by the flurry of critical articles published in the wake of the event. A headline in **Wired** captured the reaction best: "Theranos Had a Chance to Clear Its Name. Instead, It Tried to Pivot."

In an interview with the **Financial Times,** Geoffrey Baird, a professor of pathology at the University of Washington, said Holmes's presentation had included "a comically small amount of data" and had "the feel of someone putting together a last-minute term paper late at night." Other lab experts were quick to note that none of the miniLab's various components were novel. All Theranos had done was make them smaller and pack them into one box, they said.

One of the miniLab tests Holmes had showcased at the conference was for Zika, the mosquito-borne virus that had damaged the brains of thousands of newborns around the world. Theranos had applied to the FDA for emergency-use authorization of the test, billing it as the first finger-stick blood test of its kind. But in another embarrassing setback, FDA inspectors soon discovered that the company had failed to include basic patient safeguards in its study, forcing it to withdraw the application.

The possibility that Holmes might pull a rabbit out of her proverbial hat at the AACC meeting had kept Theranos's restless investors from launching a mutiny. After her appearance was panned and the Zika fiasco made headlines, one of them decided it had had enough: Partner Fund, the San Francisco hedge fund that had invested close to $100 million in the company in early 2014, sued Holmes, Balwani, and the company in Delaware's Court of Chancery, alleging that they had deceived it with "a series of lies, material misstatements, and omissions." Another set of investors led by the retired banker Robert Colman filed a separate lawsuit in federal court in San Francisco. It also alleged securities fraud and sought class-action status.

Most of the other investors opted against litigation, settling instead for a grant of extra shares in exchange for a promise not to sue. One notable exception was Rupert Murdoch. The media mogul sold his stock back to Theranos for one dollar so he could claim a big tax write-off on his other earnings. With a fortune estimated at $12 billion, Murdoch could afford to lose more than $100 million on a bad investment.

David Boies and his law firm, Boies, Schiller & Flexner, stopped doing legal work for Theranos after falling out with Holmes over how to handle the federal investigations. Another big law firm, WilmerHale, took their place. A month after Holmes's AACC appearance, Heather King returned to Boies Schiller as a partner based in its Palo Alto office. Boies left the Theranos board a few months later.

Walgreens, which had sunk a total of $140 million into Theranos, filed its own lawsuit against the company, accusing it of failing to meet the "most basic quality standards and legal requirements" of the companies' contract. "The fundamental premise of the parties' contract—like any endeavor involving human health—was to help people, and not to harm them," the drugstore chain wrote in its complaint.

After initially attempting to appeal the CMS ban, Holmes resigned herself to the inevitable and closed the California lab as well as the company's second lab in Arizona, which had used only commercial analyzers. During an inspection of the Arizona facility days before it was shuttered, CMS found a multitude of problems there too.

Under a settlement with Arizona's attorney general, Theranos subsequently agreed to pay $4.65 million into a state fund that reimbursed the 76,217 Arizonans who ordered blood tests from the company.

The number of test results Theranos voided or corrected in California and Arizona eventually reached nearly 1 million. The harm done to patients from all those faulty tests is hard to determine. Ten patients have filed lawsuits alleging consumer fraud and medical battery. One of them alleges that Theranos's blood tests failed to detect his heart disease, leading him to suffer a preventable heart attack. The suits have been consolidated into a putative class action in federal court in Arizona. Whether the plaintiffs are able to prove injury in court remains to be seen.

One thing is certain: the chances that people would have died from missed diagnoses or wrong medical treatments would have risen exponentially if the company had expanded its blood-testing services to Walgreens's 8,134 other U.S. stores as it was on the cusp of doing when **Pathology Blawg**'s Adam Clapper reached out to me.

Epilogue

In the days after my first **Journal** article, Holmes defiantly asserted that she would publish clinical data from her blood-testing system to disprove my reporting. "Data is a powerful thing because it speaks for itself," she said on October 26, 2015, at a conference hosted by the Cleveland Clinic. Two years and three months later, she finally delivered on that pledge: in January 2018, Theranos published a paper about the miniLab in the peer-reviewed scientific journal **Bioengineering and Translational Medicine**. The paper described the device's components and inner workings and included some data purporting to show that it held its own when compared with FDA-approved machines. But there was one major catch: the blood Theranos had used in its study was drawn the old-fashioned way, with a needle in the arm. Holmes's original premise—fast and accurate test results from just a drop or two pricked from a finger—was nowhere to be found in the paper.

A close read revealed other significant shortcomings. For one thing, the paper included data for only a few blood tests. And results for two of those tests, HDL cholesterol and LDL cholesterol, diverged from the FDA-approved machines by a margin that Theranos itself acknowledged "exceeds recommended limits." The company also conceded that it had run the assays one at a time, belying Holmes's previous claim that her technology could do dozens of tests simultaneously on one tiny blood sample. Last but not least, the tests performed had required different configurations of the miniLab because Theranos hadn't yet figured out how to fit all the components into one box. All of this was a far cry from the revolutionary breakthrough Holmes had touted when Theranos launched its tests in Walgreens stores in the fall of 2013.

Holmes's name was listed among the paper's coauthors but Balwani's was not. After their breakup and his departure from the company in the spring of 2016, Balwani seemed to have dropped off the face of the earth. Holmes had moved out of the 6,555-square-foot house he owned in Atherton (acquired for $9 million in 2013 through a limited liability company), and it wasn't clear if he continued to live there. For a time, there was speculation among former Theranos employees that he had fled the country to elude federal investigators.

Those rumors were put to rest on the morning of March 6, 2017, when Tyler Shultz entered a conference room at the offices of Gibson, Dunn & Crutcher on Mission Street in San Francisco. Standing among the half dozen lawyers present to take his deposition in the Partner Fund litigation was the familiar diminutive figure with the angry scowl who had terrorized Theranos employees. Balwani was a named defendant in the lawsuit, so his presence was unusual and seemed to have but one purpose: to intimidate the witness. If that was indeed the goal, it didn't work. Over the next eight and a half hours, Tyler focused on giving truthful answers to the questions he was asked and blocked out the silent presence of his irascible former boss at the other end of the conference table. Seven weeks later, Theranos settled the case for $43 million on the eve of Balwani's own deposition. (Soon after, it settled the Walgreens lawsuit for more than $25 million.)

By late 2017, Theranos was running on fumes, having burned through most of the $900 million it raised from investors, much of it on legal expenses. Several rounds of layoffs had reduced the size of its workforce to fewer than 130 employees from a high of 800 in 2015. To save on rent, the company had moved all its remaining staff to the Newark facility across San Francisco Bay. The specter of a bankruptcy filing loomed. But a few days before Christmas, Holmes announced that she had secured a $100 million loan from a private-equity firm. The financial lifeline came with strict conditions: the loan was collateralized by Theranos's patent portfolio and the company would have to meet certain product and operational milestones to get the money.

Less than three months later, the walls began closing in again: on March 14, 2018, the Securities and Exchange Commission charged Theranos, Holmes, and Balwani with conducting "an elaborate, years-long fraud." To resolve the agency's civil charges, Holmes was forced to relinquish her voting control over the company, give back a big chunk of her stock, and pay a $500,000 penalty. She also agreed to be barred from being an officer or director in a public company for ten years. Unable to reach a settlement with Balwani, the SEC sued him in federal court in California. In the meantime, the criminal investigation continued to gather steam. As of this writing, criminal indictments of both Holmes and Balwani on charges of lying to investors and federal officials seem a distinct possibility.

THE TERM "VAPORWARE" was coined in the early 1980s to describe new computer software or hardware that was announced with great fanfare only to take years to materialize, if it did at all. It was a reflection of the computer industry's tendency to play it fast and loose when it came to marketing. Microsoft, Apple, and Oracle were all accused of engaging in the practice at one point or another. Such overpromising became a defining feature of Silicon Valley. The harm done to consumers was minor, measured in frustration and deflated expectations.

By positioning Theranos as a tech company in the heart of the Valley, Holmes channeled this fake-it-until-you-make-it culture, and she went to extreme lengths to hide the fakery. Many companies in Silicon Valley make their employees sign nondisclosure agreements, but at Theranos the obsession with secrecy reached a whole different level. Employees were prohibited from putting "Theranos" on their LinkedIn profiles. Instead, they were told to write that they worked for a "private biotechnology company." Some former employees received cease-and-desist letters from Theranos lawyers for posting descriptions of their jobs at the company that were deemed too detailed. Balwani routinely monitored employees' emails and internet browser history. He also prohibited the use of Google Chrome on the theory that Google could use the web browser to spy on Theranos's R&D. Employees who worked at the office complex in Newark were discouraged from using the gym there because it might lead them to mingle with workers from other companies that leased space at the site.

In the part of the clinical lab dubbed "Normandy," partitions were erected around the Edisons so that Siemens technicians wouldn't be able to see them when they came to service the German manufacturer's machines. The partitions turned the room into a maze and blocked egress. The lab's windows were tinted, which made it nearly impossible to see in from the outside, but the company still taped sheets of opaque plastic to the inside. The doors to the corridor that led to the lab rooms, and the lab rooms themselves, were equipped with fingerprint scanners. If more than one person entered at a time, sensors set off an alarm and activated a camera that sent a photo to the security desk. As for surveillance cameras, they were everywhere. They were the kind with dark blue dome covers that kept you guessing about which way the lens was directed. All of this was ostensibly to protect trade secrets, but it's now clear that it was also a way for Holmes to cover up her lies about the state of Theranos's technology.

Hyping your product to get funding while concealing your true progress and hoping that reality will eventually catch up to the hype continues to be tolerated in the tech industry. But it's crucial to bear in mind that Theranos wasn't a tech company in the traditional sense. It was first and foremost a health-care company. Its product wasn't software but a medical device that analyzed people's blood. As Holmes herself liked to point out in media interviews and public appearances at the height of her fame, doctors base 70 percent of their treatment decisions on lab results. They rely on lab equipment to work as advertised. Otherwise, patient health is jeopardized.

So how was Holmes able to rationalize gambling with people's lives? One school of thought is that she became captive to Balwani's nefarious influence. Under this theory, Balwani was Holmes's Svengali and molded her—the innocent ingénue with big dreams—into the precocious young female startup founder that the Valley craved and that he was too old, too male, and too Indian to play himself. There's no question that Balwani was a bad influence. But to place all the blame on his shoulders is not only too convenient, it's inaccurate. Employees who saw the two interact up close describe a partnership in which Holmes, even if she was almost twenty years younger, had the last say. Moreover, Balwani didn't join Theranos until late 2009. By then, Holmes had already been misleading pharmaceutical companies for years about the readiness of her technology. And with actions that ranged from blackmailing her chief financial officer to suing ex-employees, she had displayed a pattern of ruthlessness at odds with the portrait of a well-intentioned young woman manipulated by an older man.

Holmes knew exactly what she was doing and she was firmly in control. When one former employee interviewed for a job at Theranos in the summer of 2011, he asked Holmes about the role of the company's board. She took offense at the question. "The board is just a placeholder," he recalls her saying. "I make all the decisions here." Her annoyance was so palpable that he thought he'd blown the interview. Two years later, Holmes made sure that the board would never be more than a placeholder. In December 2013, she forced through a resolution that assigned one hundred votes to every share she owned, giving her 99.7 percent of the voting rights. From that point on, the Theranos board couldn't even reach a quorum without Holmes. When he was later questioned about board deliberations in a deposition, George Shultz said, "We never took any votes at Theranos. It was pointless. Elizabeth was going to decide whatever she decided." This helps explain why the board never hired a law firm to conduct an independent investigation of what happened. At a publicly traded company, such an investigation would have been commissioned within days or weeks of the first media revelations. But at Theranos, nothing could be decided or done without Holmes's assent.

If anything, it was Holmes who was the manipulator. One after another, she wrapped people around her finger and persuaded them to do her bidding. The first to fall under her spell was Channing Robertson, the Stanford engineering professor whose reputation helped give her credibility when she was just a teenager. Then there was Donald L. Lucas, the aging venture capitalist whose backing and connections enabled her to keep raising money. Dr. J and Wade Miquelon at Walgreens and Safeway CEO Steve Burd were next, followed by James Mattis, George Shultz, and Henry Kissinger (Mattis's entanglement with Theranos proved no obstacle to his being confirmed as President Donald Trump's secretary of defense). David Boies and Rupert Murdoch complete the list, though I've left out many others who were bewitched by Holmes's mixture of charm, intelligence, and charisma.

A sociopath is often described as someone with little or no conscience. I'll leave it to the psychologists to decide whether Holmes fits the clinical profile, but there's no question that her moral compass was badly askew. I'm fairly certain she didn't initially set out to defraud investors and put patients in harm's way when she dropped out of Stanford fifteen years ago. By all accounts, she had a vision that she genuinely believed in and threw herself into realizing. But in her all-consuming quest to be the second coming of Steve Jobs amid the gold rush of the "unicorn" boom, there came a point when she stopped listening to sound advice and began to cut corners. Her ambition was voracious and it brooked no interference. If there was collateral damage on her way to riches and fame, so be it.

Acknowledgments

This book, which flowed from my work exposing the Theranos scandal in the pages of the **Wall Street Journal,** would not have been possible without the help of the confidential sources who spoke to me at great personal peril throughout 2015 and 2016. Some, like Tyler Shultz, have since gone on the record and appear under their real identities in the narrative. Others appear under pseudonyms or are mentioned only as unnamed sources. All were moved to talk to me, despite the legal and career risks they faced, by one overriding concern: protecting the patients who stood to suffer harm from Theranos's faulty blood tests. I will forever be grateful to them for their integrity and their courage. They are the true heroes of this story.

This book would also not have been possible without the dozens of other former Theranos employees who overcame their initial skittishness to share their experiences with me and help me reconstruct the company's fifteen-year history. To a person, they were generous with their time and incredibly supportive of this endeavor. I'm also indebted to the laboratory experts who schooled me in the arcane but fascinating science of blood testing. One of them, Stephen Master of Weill Cornell Medical Center in New York, was kind enough to review the manuscript before publication to help me avoid errors.

This book started with a tip in early 2015. I would like to thank the person who gave me the long leash and the unflinching support I needed to follow that tip where it led: my editor at the **Journal,** Mike Siconolfi. Mike has been a mentor, not just to me but to generations of reporters, and is a standard-bearer for the great journalistic institution that is the **Wall Street Journal**. Mike wasn't my only ally in the quest to bring these events to light: Jason Conti, now Dow Jones & Co.'s general counsel, and Jacob Goldstein, his deputy, spent countless hours vetting my reporting and batting back the legal threats made by Theranos's lawyers. I also owe a big debt of gratitude to my investigative team colleague Christopher Weaver, who helped me cover the regulatory inquiries and other fallout for more than a year, including during a stretch when I was on book leave.

One of the dividends of working at the **Journal** has been the friendships I've made there over the years. One of those friends, Christopher Stewart, has written several nonfiction books and generously shared with me both his publishing industry expertise and his contacts. It's through Chris that I met my agent, Eric Lupfer of Fletcher & Company, who immediately saw the potential of this project and pushed me to pursue it despite various obstacles that sprang up along the way. Eric's perpetual optimism was contagious and the perfect antidote for my moments of doubt.

I am very lucky that this book landed at Knopf and in the deft hands of Andrew Miller. Andrew's enthusiasm and his unshakable faith in me gave me the confidence I needed to bring it to fruition. I'm also humbled to have had the support of Andrew's boss, Knopf Doubleday Publishing Group chairman Sonny Mehta. From the moment I entered the Random House Tower, Andrew, Sonny, and their colleagues welcomed me and made me feel at home. I hope I have lived up to their expectations.

This saga has consumed the last three and a half years of my life. Through it all, I have been fortunate to be able to rely on the advice, support, and warmth of friends and family. Ianthe Dugan, Paulo Prada, Philip Shishkin, and Matthew Kaminski—to name but a few—provided frequent encouragement and much-needed comic relief. My parents, Jane and Gérard, and my sister, Alexandra, cheered me on to the finish line. But by far my greatest source of strength and inspiration came from the four people with whom I share my life: my wife, Molly, and my three children, Sebastian, Jack, and Francesca. This book is dedicated to them.

Notes

PROLOGUE

3 **"Elizabeth called me this morning":** Email with the subject line "Message from Elizabeth" sent by Tim Kemp to his team at 10:46 a.m. PST on November 17, 2006.

5 **His expert testimony:** Simon Firth, "The Not-So-Retiring Retirement of Channing Robertson," Stanford School of Engineering website, February 28, 2012.

9 **By any measure, it was a resounding success:** VC Experts report on Theranos Inc. created on December 28, 2015.

10 **A slide deck listed six deals:** PowerPoint titled "Theranos: A Presentation for Investors" dated June 1, 2006.

1. A PURPOSEFUL LIFE

12 **When she was seven:** Ken Auletta, "Blood, Simpler," **New Yorker,** December 15, 2014.

13 **On her father's side:** P. Christiaan Klieger, **The Fleischmann Yeast Family** (Charleston: Arcadia Publishing, 2004), 9.

13 **Aided by the political and business connections:** Ibid., 49.

13 **So the case could be made:** Sally Smith Hughes, interview of Donald L. Lucas for an oral history titled "Early Bay Area Venture Capitalists: Shaping the Economic and Business Landscape," Bancroft Library, University of California, Berkeley, 2010.

13 **Her father was a West Point graduate:** Obituary of George Arlington Daoust Jr., **Washington Post,** October 8, 2004.

14 **"I grew up with those stories":** Auletta, "Blood, Simpler."

17 **Midway through high school:** Ibid.

17 **Her father had drilled into her:** Roger Parloff, "This CEO Is Out for Blood," **Fortune,** June 12, 2014.

17 **The message Elizabeth took away:** Rachel Crane, "She's America's Youngest Female Billionaire—and a Dropout," CNNMoney website, October 16, 2014.

19 **The experience left her convinced:** Parloff, "This CEO Is Out for Blood."

19 **When she got back home:** Ibid.

20 **In court testimony years later: Theranos, Inc. and Elizabeth Holmes v. Fuisz Pharma LLC, Richard C. Fuisz and Joseph M. Fuisz,** No. 5:11-cv-05236-PSG, U.S. District Court in San Jose, trial transcript, March 13, 2014, 122–23.

21 **To raise the money she needed:** Sheelah Kolhatkar and Caroline Chen, "Can Elizabeth Holmes Save Her Unicorn?" **Bloomberg Businessweek,** December 10, 2015.

21 **The Draper name carried:** Danielle Sacks, "Can VCs Be Bred? Meet the New Generation in Silicon Valley's Draper Dynasty," **Fast Company,** June 14, 2012.

22 **In a twenty-six-page document:** Theranos Inc. confidential summary dated December 2004.

22 **One morning in July 2004:** John Carreyrou, "At Theranos, Many Strategies and Snags," **Wall Street Journal,** December 27, 2015.

23 **MedVenture Associates wasn't the only venture capital firm:** VC Experts report on Theranos Inc.

23 **In addition to Draper and Palmieri:** "Theranos: A Presentation for Investors," June 1, 2006.

25 **Their little enterprise:** "Stopping Bad Reactions," **Red Herring,** December 26, 2005.

25 **The email ended with:** Email with the subject line "Happy Happy Holidays" sent by Elizabeth Holmes to Theranos employees at 9:57 a.m. PST on December 25, 2005.

2. The Gluebot

29 **Having already blown through:** VC Experts report on Theranos Inc., created on December 28, 2015.

31 **Ed had noticed a quote:** Rachel Barron, "Drug Diva," **Red Herring,** December 15, 2006.

32 **Lucas and Ellison had both invested:** "Theranos: A Presentation for Investors," June 1, 2006.

33 **In Oracle's early years:** Mike Wilson, **The Difference Between God and Larry Ellison** (New York: William Morrow, 1997), 94–103.

34 **On their return to California:** Email with the subject line "Congratulations" sent by Elizabeth Holmes to Theranos employees at 11:35 a.m. PST on August 8, 2007.

38 **Theranos filed its fourteen-page complaint: Theranos Inc. v. Avidnostics Inc.,** No. 1-07-cv-093-047, California Superior Court in Santa Clara, complaint filed on August 27, 2007, 12–14.

41 **The technique was not new:** Anthony K. Campbell, "Rainbow Makers," **Chemistry World,** June 1, 2003.

3. Apple Envy

44 **In January of that year:** John Markoff, "Apple Introduces Innovative Cellphone," **New York Times,** January 9, 2007.

45 **One of them was Ana Arriola:** Ana used to be a

man named George. She transitioned from male to female after she worked at Theranos.

48 **"We have lost sight of our business objective":** Email with the subject line "IT" sent by Justin Maxwell to Ana Arriola in the early morning hours of September 20, 2007.

51 **Avie was one of Steve Jobs's oldest:** Walter Isaacson, **Steve Jobs** (New York: Simon & Schuster, 2011), 259, 300, 308.

56 **On her way out:** Email sent by Ana Arriola to Elizabeth Holmes and Tara Lencioni at 2:57 p.m. PST on November 15, 2007.

56 **Elizabeth emailed her back:** Email sent by Elizabeth Holmes to Ana Arriola at 3:27 p.m. PST on November 15, 2007.

58 **He also noticed that Elizabeth:** Email with the subject line "RE: Waiver & Resignation Letter" sent by Michael Esquivel to Avie Tevanian at 12:41 a.m. PST on December 23, 2007.

58 **At 11:17 p.m. on Christmas Eve:** Email with the subject line "RE: Waiver & Resignation Letter" sent by Michael Esquivel to Avie Tevanian at 11:17 p.m. PST on December 24, 2007.

59 **The brutal tactics used:** Letter from Avie Tevanian to Don Lucas dated December 27, 2007.

4. GOODBYE EAST PALY

74 **Its objective was to prove:** Confidential "Theranos Angiogenesis Study Report."

74 **The night before that second meeting:** John Carreyrou, "At Theranos, Many Strategies and

Snags," **Wall Street Journal,** December 27, 2015.

79 **In one of their last email exchanges:** Email with the subject line "Reading Material" sent by Justin Maxwell to Elizabeth Holmes at 7:54 p.m. PST on May 7, 2008.

79 **His resignation email read in part:** Email with the subject line "official resignation" sent by Justin Maxwell to Elizabeth Holmes at 5:19 p.m. PST on May 9, 2008.

5. THE CHILDHOOD NEIGHBOR

82 **Elizabeth's mother, Noel: Theranos, Inc. et al. v. Fuisz Pharma LLC et al.,** deposition of Lorraine Fuisz taken on June 11, 2013, in Los Angeles, 18–19.

83 **Noel and Lorraine were in and out:** Ibid., 19–20.

83 **One evening, the power went out:** Ibid., 54.

83 **Chris's grandfather:** P. Christiaan Klieger, **Moku o Lo'e: A History of Coconut Island** (Honolulu: Bishop Museum Press, 2007), 54–121.

84 **and Chris's father, Christian III:** Deposition of Lorraine Fuisz, 52.

84 **The two women stayed in regular contact:** Ibid., 22.

84 **When the Holmeses returned:** Ibid., 35.

84 **Lorraine later visited Noel:** Ibid., 23–24.

84 **On subsequent trips:** Ibid., 55–56, 100–101.

85 **Having just purchased a new house: Theranos, Inc. et al. v. Fuisz Pharma LLC et al.,** deposi-

tion of Richard Fuisz taken on June 9, 2013, in Los Angeles, 92–93.

85 **Chris and Noel Holmes did eventually: Theranos, Inc. et al. v. Fuisz Pharma LLC et al.,** deposition of Christian R. Holmes IV taken in Washington, D.C., on April 7, 2013, 30.

85 **At first, they stayed with friends:** Deposition of Lorraine Fuisz, 34.

85 **Over lunch one day:** Ibid., 65–68.

86 **When she got home:** Ibid.

86 **As he would put it years later:** Email without a subject line sent by Richard Fuisz to me at 10:57 a.m. EST on February 2, 2017.

87 **Fuisz sued Baxter:** Thomas M. Burton, "On the Defensive: Baxter Fails to Quell Questions on Its Role in the Israeli Boycott," **Wall Street Journal,** April 25, 1991.

87 **The two sides reached a settlement:** Sue Shellenbarger, "Off the Blacklist: Did Hospital Supplier Dump Its Israel Plant to Win Arabs' Favor?" **Wall Street Journal,** May 1, 1990.

88 **He sent a female operative:** Ibid.

88 **Fuisz sent one copy:** Ibid.

89 **He subsequently obtained:** Burton, "On the Defensive."

89 **In March 1993:** Thomas M. Burton, "Caught in the Act: How Baxter Got off the Arab Blacklist, and How It Got Nailed," **Wall Street Journal,** March 26, 1993.

89 **The reputational damage:** Thomas M. Burton, "Premier to Reduce Business with Baxter

to Protest Hospital Supplier's 'Ethics,'" **Wall Street Journal,** May 26, 1993.

90 **The crowning flourish:** "At Yale, Honors for an Acting Chief," **New York Times,** May 25, 1993.

90 **Three months later:** Thomas J. Lueck, "A Yale Trustee Who Was Criticized Resigns," **New York Times,** August 28, 1993.

90 **He later sold the public corporation:** "Biovail to Buy Fuisz Technologies for $154 Million," Dow Jones, July 27, 1999.

91 **In the interview, she'd described:** Interview of Elizabeth Holmes by Moira Gunn on "BioTech Nation," May 3, 2005.

91 **His thirty-five years of experience:** Deposition of Richard Fuisz, 302.

92 **"Al, Joe and I would like to patent":** Email with the subject line "Blood Analysis—deviation from norm (individualized)" sent by Richard Fuisz to Alan Schiavelli at 7:30 p.m. EST on September 23, 2005.

92 **Fuisz finally got his attention:** Email with no subject line sent by Richard Fuisz to Alan Schiavelli at 11:23 p.m. EST on January 11, 2006.

93 **Fuisz and Schiavelli exchanged more emails:** Letter dated April 24, 2006, emailed by Alan Schiavelli to Richard Fuisz advising him that the patent application had been filed, enclosing a copy of the application and a bill for services rendered.

93 **It made no secret:** Patent application no. 60794117 titled "Bodily fluid analyzer, and

system including same and method for programming same," filed on April 24, 2006, and published on January 3, 2008.

93 **The Holmeses settled into:** Deposition of Lorraine Fuisz, 32.

93 **Lorraine drove over from McLean:** Ibid., 33.

94 **She had just been profiled in Inc.:** Jasmine D. Adkins, "The Young and the Restless," **Inc.,** July 2006.

94 **It was easy for a small company:** Deposition of Richard Fuisz, 298.

94 **One dinner was at Sushiko:** Deposition of Lorraine Fuisz, 33.

94 **Chris didn't eat much:** Ibid., 33–34.

95 **Whatever the case:** Ibid., 45–46.

96 **During one encounter:** Ibid., 42.

96 **As she was cutting Lorraine's hair:** Ibid., 40–41.

96 **Lorraine Fuisz and Noel Holmes saw each other:** Ibid., 108–10.

97 **However, Theranos didn't learn:** Email with the subject line "Is this something new?" sent by Gary Frenzel to Elizabeth Holmes, Ian Gibbons, and Tony Nugent at 11:53 p.m. PST on May 14, 2008.

98 **She came by a few weeks later: Theranos, Inc. et al. v. Fuisz Pharma LLC et al.,** declaration of Charles R. Work executed in Stevensville, Maryland, on July 22, 2013.

99 **Elizabeth got straight to the point:** Ibid.

100 **He informed her of his decision:** Ibid.

6. SUNNY

104 **Sunny had been a presence:** Ken Auletta, "Blood, Simpler," **New Yorker,** December 15, 2014.

104 **Elizabeth had struggled to make friends: Theranos, Inc. et al. v. Fuisz Pharma LLC et al.,** deposition of Lorraine Fuisz, 85–86.

104 **Born and raised in Mumbai:** LinkedIn profile of Sunny Balwani; Theranos website.

105 **Analysts were breathlessly predicting:** Steve Hamm, "Online Extra: From Hot to Scorched at Commerce One," **Bloomberg Businessweek,** February 3, 2003.

105 **It finished the year up:** Ibid.

105 **That November:** "Commerce One Buys CommerceBid for Stock and Cash," **New York Times,** November 6, 1999.

105 **It was a breathtaking price:** "Commerce One to Buy CommerceBid," CNET website, November 6, 1999.

105 **Commerce One eventually filed:** Eric Lai, "Commerce One Rises from Dot-Ashes," **San Francisco Business Times,** March 3, 2005.

106 **When they'd first met in China:** Deed for a property at the corner of Marina Boulevard and Scott Street in San Francisco, dated March 2, 2001, listing Sunny Balwani and Keiko Fujimoto as husband and wife.

106 **By October 2004:** Deed for 325 Channing Avenue #118, Palo Alto, California 94301, dated October 29, 2004.

106 **Other public records:** The TLO records-search service lists Elizabeth Holmes as residing at 325 Channing Avenue #118 in Palo Alto beginning in July 2005. In her voter registration form dated October 10, 2006, she also listed that address as her residence.

106 **He had stayed on at Commerce One:** LinkedIn profile of Sunny Balwani; Theranos website.

106 **The maneuver generated an artificial tax loss: Ramesh Balwani v. BDO Seidman, L.L.P. and François Hechinger,** No. CGC-04-433732, California Superior Court in San Francisco, complaint filed on August 11, 2004, 10.

107 **He turned around and sued BDO: Ramesh Balwani v. BDO Seidman et al.,** 4, 6–7.

111 **Elizabeth had tried to put the best spin:** Confidential "Theranos Angiogenesis Study Report."

7. DR. J

123 **In June 2010, the social network's:** Alexei Oreskovic, "Elevation Partners Buys $120 million in Facebook Shares," Reuters, June 28, 2010.

123 **Six months later:** Susanne Craig and Andrew Ross Sorkin, "Goldman Offering Clients a Chance to Invest in Facebook," **New York Times,** January 2, 2011.

123 **The emergence of Twitter:** Michael Arrington, "Twitter Closing New Venture Round at $1 Billion Valuation," TechCrunch website, September 16, 2009.

123 **In the spring of 2010:** Christine Lagorio-Chafkin,

"How Uber Is Going to Hire 1,000 People This Year," **Inc.**, January 15, 2014.

124 **Dr. J operated out of an office:** LinkedIn profile of Jay Rosan; Jessica Wohl, "Walgreen to Buy Clinic Operator Take Care Health," Reuters, May 16, 2007.

124 **In January 2010, Theranos had approached Walgreens: Walgreen Co. v. Theranos, Inc.**, No. 1:16-cv-01040-SLR, U.S. District Court in Wilmington, complaint filed on November 8, 2016, 4–5.

125 **Two months later, Elizabeth and Sunny:** Ibid., 5–6.

127 **On the Walgreens side:** Minutes of August 24, 2010, meeting between Walgreens and Theranos.

127 **"I'm so excited that we're doing this!":** Ibid.

127 **It would involve:** Schedule F of Theranos Master Purchase Agreement dated July 30, 2010, filed as Exhibit C in **Walgreen Co. v. Theranos, Inc.** complaint.

128 **A preliminary contract:** Schedule B, F, and H1 of July 2010 Theranos Master Purchase Agreement.

129 **Theranos had told Walgreens:** Document with a Theranos logo titled "Theranos Base Assay Library."

131 **When the Walgreens side had broached:** Confidential memo titled "WAG / Theranos site visit thoughts and Recommendations" addressed by Kevin Hunter to Walgreens executives on August 26, 2010.

132 **Standing in front of a slide:** PowerPoint titled "Project Beta—Disrupting the Lab Industry—Kickoff Review" dated September 28, 2010.

133 **In a report he'd put together:** Hunter's August 26, 2010, memo to Walgreens executives.

133 **Hunter asked about the blood-test results:** Minutes of video conference between Theranos and Walgreens held between 1:00 p.m. and 2:00 p.m. CDT on October 6, 2010.

134 **Elizabeth and Sunny had a testy exchange:** Minutes of video conference between Theranos and Walgreens held between 1:00 p.m. and 2:00 p.m. CDT on November 10, 2010.

136 **The contract the companies had signed**: Schedule B of July 2010 Theranos Master Purchase Agreement.

136 **Documents it gave Walgreens stated:** "Project Beta—Disrupting the Lab Industry—Kickoff Review," 5.

137 **It was a letter dated April 27, 2010:** Letter marked confidential on Johns Hopkins Medicine letterhead titled "Summary of Hopkins/Walgreens/Theranos" meeting.

138 **He'd gotten hooked on the subject:** Richard S. Dunham and Keith Epstein, "One CEO's Health-Care Crusade," **Bloomberg Businessweek,** July 3, 2007.

138 **He'd pioneered innovative wellness:** Jaime Fuller, "Barack Obama and Safeway: A Love Story," **Washington Post,** February 18, 2014.

138 **Like Dr. J, he was serious:** Dunham and Epstein, "One CEO's Health-Care Crusade."

141 **However, many of his colleagues:** Melissa Harris and Brian Cox, "2nd DUI Arrest for Walgreen Co. CFO Wade Miquelon," **Chicago Tribune,** October 18, 2010.

8. THE MINILAB

146 **The first commercial spectrophotometer:** Jerry Gallwas, "Arnold Orville Beckman (1900–2004)," **Analytical Chemistry,** August 1, 2004, 264A–65A.

146 **Cytometry, a way of counting blood cells:** M. L. Verso, "The Evolution of Blood-Counting Techniques," **Medical History** 8, no. 2 (April 1964): 149–58.

146 **One of them, a device:** Abaxis brochure for the "Piccolo Xpress chemistry analyzer" available on the Abaxis website.

9. THE WELLNESS PLAY

165 **The supermarket chain had just announced:** Safeway, "Safeway Inc. Announces Fourth Quarter 2011 Results," press release, February 23, 2012.

165 **One of them, Ed Kelly:** Conference call on Safeway's fourth-quarter 2011 earnings held at 11:00 a.m. EST on February 23, 2012, available on Earningscast.com.

166 **Piqued, Burd said he disagreed:** Ibid.

172 **A few months earlier:** CMS Form 2567 indicating an inspection of Theranos's laboratory at 3200 Hillview Avenue in Palo Alto was completed on January 9, 2012, with no deficiencies found.

172 **Although the ultimate enforcer:** California Bureau of State Audits, "Department of Public Health: Laboratory Field Services' Lack of Clinical Laboratory Oversight Places the Public at Risk," September 2008.

173 **To Dupuy, Lim's blunders were inexcusable:** Letter dated June 25, 2012, sent by attorney Jacob Sider to Elizabeth Holmes on behalf of Diana Dupuy.

174 **The phlebotomists hadn't been trained to use:** Ibid.

176 **The email, on which she copied Elizabeth:** Email with the subject line "Events" sent by Diana Dupuy to Sunny Balwani, copying Elizabeth Holmes, at 11:13 a.m. PST on May 27, 2012.

177 **Sunny agreed to have someone:** Email with the subject line "RE: Observations" sent by Sunny Balwani to Diana Dupuy, copying Elizabeth Holmes, at 2:16 p.m. PST on May 27, 2012.

177 **Over the next several days:** Emails with the subject lines "Important notice from Theranos" and "RE: Important notice from Theranos" sent by David Doyle to Diana Dupuy on May 29, May 30, and June 1, 2012.

177 **Dupuy initially refused:** Sider's June 25, 2012, letter to Holmes.

177 **Burd was asked about the status:** Conference call on Safeway's first-quarter 2012 earnings held at 11:00 a.m. EST on April 26, 2012, available on Earningscast.com.

178 **In the next earnings call:** Conference call on Safeway's second-quarter 2012 earnings held at 11:00 a.m. EST on July 19, 2012, available on Earningscast.com.

179 **Shortly after the stock market closed:** Safeway, "Safeway Announces Retirement of Chairman and CEO Steve Burd," press release, January 2, 2013.

179 **Among a list of his achievements:** Ibid.

180 **Just three months after leaving:** "Letter from Steve Burd, Founder and CEO" at Burdhealth.com.

10. "WHO IS LTC SHOEMAKER?"

182 **The idea of using Theranos devices:** Carolyn Y. Johnson, "Trump's Pick for Defense Secretary Went to the Mat for the Troubled Blood-Testing Company Theranos," **Washington Post,** December 1, 2016.

188 **With the approval of his boss:** Email with the subject line "Seeking regulatory advice regarding Theranos (UNCLASSIFIED)" sent by David Shoemaker to Sally Hojvat at 10:16 a.m. EST on June 14, 2012.

188 **Hojvat forwarded his query:** Email with the

subject line "FW: Seeking regulatory advice regarding Theranos (UNCLASSIFIED)" sent by Sally Hojvat to Elizabeth Mansfield, Katherine Serrano, Courtney Lias, Alberto Gutierrez, Don St. Pierre, and David Shoemaker at 11:43 a.m. EST on June 15, 2012.

188 **However, in practice, it had not done so:** Office of Public Health Strategy and Analysis, Office of the Commissioner, Food and Drug Administration, "The Public Health Evidence for FDA Oversight of Laboratory Developed Tests: 20 Case Studies," November 16, 2015.

189 **That changed in the 1990s:** Ibid.

189 **Gutierrez forwarded the Shoemaker email:** Email with the subject line "FW: Seeking regulatory advice regarding Theranos (UNCLASSIFIED)" sent by Alberto Gutierrez to Judith Yost, Penny Keller, and Elizabeth Mansfield at 4:36 p.m. EST on July 15, 2012.

190 **Yost and Keller decided it wouldn't hurt:** Email with thc subject line "RE: Seeking regulatory advice regarding Theranos (UNCLASSIFIED)" sent by Judith Yost to Penny Keller and Sarah Bennett at 11:46 a.m. EST on June 18, 2012.

190 **The job fell to Gary Yamamoto:** Email with the subject line "FW: Seeking regulatory advice regarding Theranos (UNCLASSIFIED)" sent by Penny Keller to Gary Yamamoto at 5:48 p.m. EST on June 18, 2012.

190 **Two months later, on August 13, 2012:** Email

with the subject line "RE: Theranos update?" sent by Gary Yamamoto to Penny Keller and Karen Fuller at 2:03 p.m. EST on August 15, 2012.

190 **When he explained that his agency:** Email with the subject line "RE: Theranos (UNCLASSIFIED)" sent by Penny Keller to David Shoemaker, copying Erin Edgar, at 1:36 p.m. EST on August 16, 2012.

192 **In a blistering email to General Mattis:** Email with the subject line "RE: Follow up" sent by Elizabeth Holmes to James Mattis, copying Jorn Pung and Karl Horst, at 3:14 p.m. EST on August 9, 2012.

193 **He forwarded it to Colonel Erin Edgar:** Email with the subject line "FW: Follow up" sent by James Mattis to Erin Edgar, copying Karl Horst, Carl Mundy, and Jorn Pung, at 10:52 p.m. EST on August 9, 2012.

194 **He also forwarded to Shoemaker:** Email with the subject line "Fw: Follow up" sent by Erin Edgar to David Shoemaker at 1:35 p.m. EST on August 14, 2012.

194 **The blunt-spoken general had:** Thomas E. Ricks, **Fiasco** (New York: The Penguin Press, 2006), 313.

195 **With Colonel Edgar's encouragement:** Email with the subject line "Theranos (UNCLASSIFIED)" sent by David Shoemaker to Penny Keller and Judith Yost, copying Erin Edgar and Robert Miller, at 3:34 p.m. EST on August 15, 2012.

195 **The response he got:** Email with the subject line "RE: Theranos (UNCLASSIFIED)" sent by Penny Keller to David Shoemaker, copying Erin Edgar, at 1:36 p.m. EST on August 16, 2012.

195 **When he confronted Colonel Edgar:** Email with the subject line "Re: Theranos (UNCLASSIFIED)" sent by Erin Edgar to David Shoemaker at 7:23 p.m. EST on August 16, 2012.

196 **At 3:00 p.m. sharp on August 23, 2012:** Email with the subject line "RE: Theranos followup (UNCLASSIFIED)" sent by David Shoemaker to Alberto Gutierrez at 10:58 a.m. EST on August 20, 2012.

11. LIGHTING A FUISZ

199 **The doorbell at 1238 Coldwater Canyon Drive:** Affidavit of service of summons notarized on October 31, 2011.

199 **The couple had purchased it: Theranos, Inc. et al. v. Fuisz Pharma LLC et al.,** deposition of Lorraine Fuisz, June 11, 2013, 111; Realtor.com.

200 **He had sold it:** "Biovail to Buy Fuisz Technologies for $154 Million," Dow Jones, July 27, 1999.

200 **It was now part of:** "Biovail to Merge with Valeant," **New York Times,** June 21, 2010.

200 **The lawsuit had been filed: Theranos, Inc. et al. v. Fuisz Pharma LLC et al.,** complaint filed on October 26, 2011, 7–10.

201 **The first and only time Fuisz:** Email with the subject line "http://www.freshpatents.com/

Medical-device-for-analyte-monitoring-and-drug-delivery-dt20060323ptan20060062852.php" sent by Richard Fuisz to John Fuisz, copying Joe Fuisz, at 8:31 a.m. EST on July 3, 2006.

201 **John replied that McDermott:** Email with the subject line "Re: http://www.freshpatents.com/Medical-device-for-analyte-monitoring-and-drug-delivery-dt20060323ptan20060062852.php" sent by John Fuisz to Richard Fuisz, copying Joe Fuisz, at 9:34 a.m. EST on July 3, 2006.

201 **John had no reason to wish: Theranos, Inc. et al. v. Fuisz Pharma LLC et al.,** deposition of Lorraine Fuisz, 80–81, 83.

202 **Noel had even dropped by: Theranos, Inc. et al. v. Fuisz Pharma LLC et al.,** deposition of John Fuisz taken on May 29, 2013, in Washington, D.C., 38.

202 **Fuisz had rubbed that fact:** Email with the subject line "Gen Dis" sent by Richard Fuisz to info@theranos.com at 7:29 a.m. PST on November 8, 2010.

203 **On his way to a resounding:** David Margolick, "The Man Who Ate Microsoft," **Vanity Fair,** March 1, 2000.

203 **In one case that illustrated:** John R. Wilke, "Boies Will Be Boies, as Another Legal Saga in Florida Shows," **Wall Street Journal,** December 6, 2000.

203 **After a judge in Miami:** Ibid.

205 **One of them was a declaration: Theranos, Inc.**

et al. v. Fuisz Pharma LLC et al., declaration of Brian B. McCauley executed in Washington, D.C., on January 12, 2012.

205 **But in a response five days later:** Letter dated January 17, 2012, sent by David Boies to Elliot Peters.

206 **He also offered to meet:** Letter dated June 7, 2012, sent by Richard Fuisz to Donald L. Lucas, Channing Robertson, T. Peter Thomas, Robert Shapiro, and George Shultz.

206 **The only response he got:** Letter dated July 5, 2012, sent by David Boies to Jennifer Ishimoto.

206 **In 1992, when John was fresh: Terex Corporation et al. v. Richard Fuisz et al.,** No. 1:1992-cv-0941, U.S. District Court for the District of Columbia, deposition of John Fuisz taken on February 17, 1993, in Washington, D.C., 118–54.

206 **At the time, Richard Fuisz:** "Manufacturer Sues Seymour Hersh over Scud Launcher Report," Associated Press, April 17, 1992.

207 **Even though the incident was twenty years old: Terex Corporation et al. v. Richard Fuisz et al.,** stipulation filed on December 2, 1996, by Judge Royce C. Lamberth dismissing case with prejudice.

207 **Boies's strategy of painting John: Theranos, Inc. et al. v. Fuisz Pharma LLC et al.,** order filed on June 6, 2012, granting defendant John R. Fuisz's motion to dismiss and granting in part and denying in part Fuisz Pharma LLC,

Richard C. Fuisz, and Joseph M. Fuisz's motion to dismiss.

208 **Boies turned around and sued: Theranos, Inc. et al. v. McDermott, Will & Emery LLP,** No. 2012-CA-009617-M, Superior Court of the District of Columbia, complaint filed on December 29, 2012.

208 **"Simply because attorneys": Theranos, Inc. et al. v. McDermott, Will & Emery LLP,** order filed on August 2, 2013, granting defendant McDermott's motion to dismiss with prejudice.

209 **Asked by one of his father's lawyers: Theranos, Inc. et al. v. Fuisz Pharma LLC et al.,** deposition of John Fuisz, 238.

210 **Boies charged clients:** Vanessa O'Connell, "Big Law's $1,000-Plus an Hour Club," **Wall Street Journal,** February 23, 2011; David A. Kaplan, "David Boies: Corporate America's No. 1 Hired Gun," **Fortune,** October 20, 2010.

212 **But then something strange happened: Theranos, Inc. et al. v. Fuisz Pharma LLC et al.,** transcript of pretrial conference and hearing on motions, March 5, 2014, 42.

12. IAN GIBBONS

213 **Ian and Robertson had met:** U.S. Patent no. 4,946,795 issued August 7, 1990.

216 **He complained to his old friend: Theranos, Inc. et al. v. Fuisz Pharma LLC et al.,** transcript of pretrial conference and hearing on motions, March 5, 2014, 47–48.

223 **After trying for weeks: Theranos, Inc. et al. v. Fuisz Pharma LLC et al.,** defendants' notice of deposition for Ian Gibbons, filed on May 6, 2013.

223 **With the deadline for his appearance:** Email with the subject line "Deposition—Confidential A/C Privileged" sent by David Doyle to Ian Gibbons, copying Mona Ramamurthy, at 7:32 p.m. PST on May 15, 2013.

223 **Ian forwarded the email:** Email with the subject line "Fwd: FW: Deposition—Confidential A/C Privileged" sent by Ian Gibbons to Rochelle Gibbons at 7:49 p.m. PST on May 15, 2013.

13. CHIAT\DAY

228 **She'd even tried to convince Lee Clow:** Walter Isaacson, **Steve Jobs** (New York: Simon & Schuster, 2011), 162, 327.

232 **Elizabeth believed in the Flower of Life:** April Holloway, "What Ancient Secrets Lie Within the Flower of Life?" **Ancient Origins,** December 1, 2013.

235 **In an email to Kate listing items:** Email with the subject line "Legal" sent by Mike Peditto to Kate Wolff at 4:27 p.m. PST on January 4, 2013.

238 **It indemnified Chiat\Day:** Agency agreement between TBWA\CHIAT\DAY, Los Angeles and Theranos Inc. dated October 12, 2012.

238 **He fired off an email to Joe Sena:** Email with the subject line "Fwd: Contract" sent by Mike

Peditto to Joseph Sena at 6:23 p.m. PST on March 19, 2013.

238 **Sena replied:** Email with the subject line "RE: Contract" sent by Joseph Sena to Mike Peditto at 6:51 p.m. PST on March 20, 2013.

240 **But Kate and Mike stayed alert:** Many of the last-minute changes to the Theranos website are captured in a Microsoft Word document marked "Theranos Confidential" that Jeff Blickman emailed to Kate Wolff and Mike Peditto moments before the conference call.

14. GOING LIVE

245 **He had just finished reading:** Walter Isaacson, **Steve Jobs** (New York: Simon & Schuster, 2011).

248 **Another was Chinmay Pangarkar:** LinkedIn profile of Chinmay Pangarkar.

248 **There was also Suraj Saksena:** LinkedIn profile of Suraj Saksena.

252 **This Frankenstein machine:** See the definition of "blade server" in the **PC Magazine Encyclopedia** available at PCMag.com.

254 **on June 5, 2012, she'd signed:** Amended and restated Theranos Master Services Agreement dated June 5, 2012, filed as Exhibit A in **Walgreen Co. v. Theranos, Inc.,** complaint.

255 **The ADVIA was a hulking:** See the Technical Specifications tab on the page devoted to the ADVIA 1800 Chemistry System on the U.S. website of Siemens Healthineers.

258 **Hemolysis was a known side effect:** Marlies Oostendorp, Wouter W. van Solinge, and Hans Kemperman, "Potassium but Not Lactate Dehydrogenase Elevation Due to In Vitro Hemolysis Is Higher in Capillary Than in Venous Blood Samples," **Archives of Pathology & Laboratory Medicine** 136 (October 2012): 1262–65.

15. UNICORN

263 **She hated the artist's illustration:** Joseph Rago, "Elizabeth Holmes: The Breakthrough of Instant Diagnosis," **Wall Street Journal,** September 7, 2013.

264 **A press release was due:** Theranos, "Theranos Selects Walgreens as a Long-Term Partner Through Which to Offer Its New Clinical Laboratory Service," press release, September 9, 2013, Theranos website.

264 **The former statesman: Theranos, Inc. et al. v. Fuisz Pharma LLC et al.,** trial transcript, March 13, 2014, 92.

265 **Nor did Rago:** "**WSJ**'s Rago Wins Pulitzer Prize," **Wall Street Journal,** April 19, 2011.

268 **A few weeks later:** Email with the subject line "Theranos-time sensitive" sent by Donald A. Lucas to Mike Barsanti and other Lucas Venture Group clients at 2:47 p.m. PST on September 9, 2013.

269 **They ranged from Robert Colman: Robert Colman and Hilary Taubman-Dye, Indi-**

vidually and on Behalf of All Others Similarly Situated, v. Theranos, Inc., Elizabeth Holmes, and Ramesh Balwani, No. 5:16-cv-06822, U.S. District Court in San Francisco, complaint filed on November 28, 2016, 4.

270 **In an article published:** Aileen Lee, "Welcome to the Unicorn Club: Learning from Billion-Dollar Startups," TechCrunch website, November 2, 2013.

270 **A few weeks before Elizabeth's:** Tomio Geron, "Uber Confirms $258 Million from Google Ventures, TPG, Looks to On-Demand Future," Forbes.com, August 23, 2013.

270 **There was also Spotify:** John D. Stoll, Evelyn Rusli, and Sven Grundberg, "Spotify Hits a High Note: Valuation Tops $4 Billion," **Wall Street Journal,** November 21, 2013.

271 **With about $4 billion in assets:** Cliffwater LLC, "Hedge Fund Investment Due Diligence Report: Partner Fund Management LP," December 2011, 2.

271 **After they reached out to her: Partner Investments, L.P., PFM Healthcare Master Fund, L.P., PFM Healthcare Principals Fund, L.P. v. Theranos, Inc., Elizabeth Holmes, Ramesh Balwani and Does 1–10,** No. 12816-VCL, Delaware Chancery Court, complaint filed on October 10, 2016, 10.

272 **During that first meeting:** Ibid., 11.

272 **At a second meeting three weeks later:** Ibid., 15–16.

273 **The rub was that much of the data: Partner Investments, L.P. et al. v. Theranos, Inc. et al.,** deposition of Pranav Patel taken on March 9, 2017, in Palo Alto, California, 95–97.

273 **Sunny also told James and Grossman: Partner Investments, L.P. et al. v. Theranos, Inc. et al.** complaint, 16–17.

273 **Sunny and Elizabeth's boldest claim:** Ibid., 12–13.

275 **A spreadsheet with financial projections: Partner Investments, L.P. et al. v. Theranos, Inc. et al.,** deposition of Danise Yam taken on March 16, 2017, in Palo Alto, California, 154–58.

276 **Six weeks after Sunny sent Partner Fund:** Ibid., 140–58.

276 **As it would turn out:** Christopher Weaver, "Theranos Had $200 Million in Cash Left at Year-End," **Wall Street Journal,** February 16, 2017.

276 **On February 4, 2014, Partner Fund: Partner Investments, L.P. et al. v. Theranos, Inc. et al.** complaint, 17–18.

16. THE GRANDSON

282 **It was as if you flipped a coin enough times: Partner Investments, L.P., PFM Healthcare Master Fund, L.P., PFM Healthcare Principals Fund, L.P. v. Theranos, Inc., Elizabeth Holmes, Ramesh Balwani and Does 1-10,** No. 12816-VCL, Delaware Chancery Court,

deposition of Tyler Shultz taken on March 6, 2017, in San Francisco, California, 138.

283 **Over a period of several days:** Email with the subject line "RE: Follow up to previous discussion" sent by Tyler Shultz to Elizabeth Holmes at 3:38 p.m. PST on April 11, 2014.

285 **Moreover, Do wasn't even authorized: Partner Investments, L.P. et al. v. Theranos, Inc. et al.,** deposition of Erika Cheung taken on March 7, 2017, in Los Angeles, California, 45–47.

286 **The inspector spent several hours:** CMS Form 2567 indicating that relatively minor deficiencies were found during an inspection of Theranos's laboratory on December 3, 2013.

292 **It could be widened at will:** Tyler Shultz's April 11, 2014, email to Elizabeth Holmes.

292 **One of them was Elizabeth's:** Joseph Rago, "Elizabeth Holmes: The Breakthrough of Instant Diagnosis," **Wall Street Journal,** September 7, 2013.

294 **Tyler had looked up the CLIA regulations:** Title 42 of the Code of Federal Regulations, Part 493, Subpart H, Section 801.

294 **At 9:16 a.m. on Monday:** Email with the subject line "RE: Proficiency Testing Question" sent by Stephanie Shulman to Colin Ramirez, aka Tyler Shultz, at 12:16 p.m. EST on March 31, 2014.

295 **In response to a description he gave her:** Email with the subject line "RE: Proficiency Testing Question" sent by Stephanie Shulman to Colin

Ramirez, aka Tyler Shultz, at 4:46 p.m. EST on April 2, 2014.

296 **So he went ahead and typed up:** Tyler Shultz's April 11, 2014, email to Elizabeth Holmes.

297 **In a point-by-point rebuttal:** Email sent by Sunny Balwani to Tyler Shultz on April 15, 2014.

302 **It said she disagreed with running:** Resignation letter written by Erika Cheung dated April 16, 2014.

17. FAME

304 **But the judge overseeing the case: Theranos, Inc. et al. v. Fuisz Pharma LLC et al.,** transcript of pretrial conference and hearing on motions, March 5, 2014, 48.

304 **One of them was Fuisz's contention: Theranos, Inc. et al. v. Fuisz Pharma LLC et al.,** trial transcript, March 14, 2014, 118–21.

304 **In his rambling opening argument: Theranos, Inc. et al. v. Fuisz Pharma LLC et al.,** trial transcript, March 13, 2014, 54.

306 **Underhill had left McDermott: Theranos, Inc. et al. v. Fuisz Pharma LLC et al.,** deposition of John Fuisz, 165–66.

308 **The next morning, Fuisz jotted down:** Handwritten note dated March 17, 2014, on Fairmont Hotels and Resorts stationery.

309 **In his pique, John emailed:** Email with the subject line "Theranos" sent by John Fuisz to Julia Love at 7:15 a.m. EST on March 17, 2014.

309 **He then forwarded the email:** Email with the subject line "Fwd: Theranos" sent by John Fuisz to Richard Fuisz, Joe Fuisz, Michael Underhill, and Rhonda Anderson at 7:17 a.m. EST on March 17, 2014.

309 **Underhill responded angrily:** Email with the subject line "RE: Theranos" sent by Michael Underhill to John Fuisz, copying David Boies, Richard Fuisz, Joe Fuisz, and Rhonda Anderson, at 3:59 p.m. EST on March 17, 2014.

309 **In case the message wasn't clear:** Email with the subject line "Re: Theranos" sent by David Boies to John Fuisz, copying Julia Love, Michael Underhill, Richard Fuisz, and Joe Fuisz, at 4:16 p.m. EST on March 17, 2014.

309 **Julia Love's article:** Julia Love, "Family Gives Up Disputed Patent, Ending Trial with Boies' Client," **Litigation Daily,** March 17, 2014.

313 **When Parloff's cover story:** Roger Parloff, "This CEO Is Out for Blood," **Fortune,** June 12, 2014.

314 **Had Parloff read Robertson's testimony: Theranos, Inc. et al. v. Fuisz Pharma LLC et al.**, trial transcript, March 14, 2014, 202.

314 **Under the headline "Bloody Amazing":** Matthew Herper, "Bloody Amazing," Forbes.com, July 2, 2014.

314 **Two months later, she graced:** "The Forbes 400," **Forbes,** October 20, 2014.

315 **Elizabeth became the youngest person:** Press release from the Horatio Alger Association on PRNewswire, March 9, 2015.

315 **Time magazine: Time,** "The 100 Most Influential People," April 16, 2015.

315 **President Obama appointed her:** Theranos, "Elizabeth Holmes on Joining the Presidential Ambassadors for Global Entrepreneurship (PAGE) Initiative," press release, May 11, 2015, Theranos website.

316 **Elizabeth also had a personal chef:** Ken Auletta, "Blood, Simpler," **New Yorker,** December 15, 2014.

316 **In September 2014, three months after:** Holmes's TEDMED speech can be viewed on YouTube; https://www.youtube.com/watch?v=kZTfgXYjj-A.

18. THE HIPPOCRATIC OATH

323 **He did send them one of his email exchanges:** Email with the subject line "Re: The Employment Law Group: Consultation Information" sent to DeWayne Scott at 9:18 p.m. EST on October 29, 2014.

329 **Phyllis and her husband:** Phyllis Gardner is listed as a scientific and strategic adviser in the confidential Theranos Inc. summary dated December 2004 that Holmes used to pitch investors during the company's Series A funding round.

331 **That would change when:** Ken Auletta, "Blood, Simpler," **New Yorker,** December 15, 2014.

331 **Among the arguments she marshaled:** Steven M. Chan, John Chadwick, Daniel L. Young, Elizabeth Holmes, and Jason Gotlib, "Intensive

Serial Biomarker Profiling for the Prediction of Neutropenic Fever in Patients with Hematologic Malignancies Undergoing Chemotherapy: A Pilot Study," **Hematology Reports** 6 (2014): 5466.

332 **In a post on his blog:** Clapper's blog post can be viewed by entering "PathologyBlawg.com" into the Wayback Machine.

19. THE TIP

336 **He'd patiently explained to me:** John Carreyrou and Janet Adamy, "How Medicare 'Self-Referral' Thrives on Loophole," **Wall Street Journal,** October 22, 2014.

337 **"A chemistry is performed so that":** Ken Auletta, "Blood, Simpler," **New Yorker,** December 15, 2014.

338 **Sure, Mark Zuckerberg had learned:** Jose Antonio Vargas, "The Face of Facebook," **New Yorker,** September 20, 2010.

338 **There was a reason many Nobel laureates:** "Average Age for Nobel Laureates in Physiology or Medicine," Nobelprize.org.

338 **In the meantime, I did some preliminary research:** Joseph Rago, "Elizabeth Holmes: The Breakthrough of Instant Diagnosis," **Wall Street Journal,** September 7, 2013.

344 **It was the last Saturday:** N. R. Kleinfield, "With White-Knuckle Grip, February's Cold Clings to New York," **New York Times,** February 27, 2015.

352 **She had written Theranos a letter:** Letter written by Dr. Sundene dated January 20, 2015, and addressed to "Theranos Quality Control."

355 **As I was wrapping up my trip:** Email with the subject line "Theranos" sent by Matthew Traub to John Carreyrou at 1:11 p.m. EST on April 21, 2015.

356 **I wrote Traub back to confirm:** Email with the subject line "Re: Theranos" sent by John Carreyrou to Matthew Traub at 7:08 p.m. EST on April 21, 2015.

356 **He said he would check:** Email with the subject line "Re: Theranos" sent by Matthew Traub to John Carreyrou at 12:02 a.m. EST on April 22, 2015.

356 **As I scanned my results:** My test results from Theranos and LabCorp were faxed to Dr. Sundene on April 24, 2015. I got my blood drawn at a Theranos wellness center in Phoenix on April 23, 2015, forty-four minutes before getting my blood drawn a second time at a LabCorp site.

356 **Those differences were mild compared:** Dr. Sundene received her test results from LabCorp on April 28, 2015, and her results from Theranos on April 30, 2015. She got her blood drawn at a LabCorp site on April 24, 2015, fifty-three minutes before getting her blood drawn a second time at a Theranos wellness center.

359 **The awkward dinner conversation:** John Carreyrou, "Theranos Whistleblower Shook

the Company—and His Family," **Wall Street Journal,** November 18, 2016.

360 **The time stamp on the attorney's email:** Email with the subject line "Deposition—Confidential A/C Privileged" sent by David Doyle to Ian Gibbons, copying Mona Ramamurthy, at 7:32 p.m. PST on May 15, 2013.

20. THE AMBUSH

363 **I had sent him an email outlining:** Email with the subject line "list of questions for Theranos" sent by John Carreyrou to Matthew Traub at 6:33 p.m. EST on June 9, 2015.

364 **Tyler arrived at his grandfather's house:** An abridged account of Tyler Shultz's ordeal was published in John Carreyrou, "Theranos Whistleblower Shook the Company—and His Family," **Wall Street Journal,** November 18, 2016.

375 **She had recently appeared:** Holmes's interviews on **CBS This Morning** (April 16, 2015), CNBC's **Mad Money** (April 27, 2015), CNN's **Fareed Zakaria GPS** (May 18, 2015), and PBS's **Charlie Rose** (June 3, 2015) can all be viewed on YouTube.

21. TRADE SECRETS

376 **Rounding out the group:** Fritsch's firm, Fusion GPS, would later gain notoriety for commissioning the infamous dossier on President Don-

ald Trump from a former British spy alleging that Trump was vulnerable to Russian blackmail.

377 **The tone was set from the start:** I also recorded the meeting. The quotes are transcribed verbatim from that recording.

377 **At Traub's request, I had sent:** Email with the subject line "list of questions for Theranos" sent by John Carreyrou to Matthew Traub at 6:33 p.m. EST on June 9, 2015.

384 **The letter inside the envelope:** Letter from David Boies to Erika Cheung dated June 26, 2015.

385 **Attached to it was a formal letter:** Letter from David Boies to Jason P. Conti, copying John Carreyrou and Mike Siconolfi, dated June 26, 2015.

386 **The next day, I received:** Email with the subject line "Re: Theranos HIPAA waiver" sent by Nicole Sundene to John Carreyrou at 7:04 p.m. EST on June 30, 2015.

387 **I sent Heather King an email:** Email with the subject line "Eric Nelson" sent by John Carreyrou to Heather King at 1:07 p.m. EST on July 1, 2015.

387 **Later that week, Boies:** Letter from David Boies to Jason P. Conti, copying Mark H. Jackson, John Carreyrou, and Mike Siconolfi, dated July 3, 2015.

387 **His main evidence to back up:** The signed statements by Drs. Rezaie and Beardsley are dated July 1, 2015.

388 **Dr. Stewart emailed a few days later:** Email

with the subject line "Theranos" sent by Dr. Stewart to John Carreyrou at 8:26 p.m. EST on July 8, 2015.

22. LA MATTANZA

390 **The first was that the FDA:** Theranos, "Theranos Receives FDA Clearance and Review and Validation of Revolutionary Finger Stick Technology, Test, and Associated System," press release, July 2, 2015, Theranos website.

390 **The second was that a new law:** Ken Alltucker, "Do-It-Yourself Lab Testing Without Doc's Orders Begins," **Arizona Republic,** July 7, 2015.

392 **The latest had been a state dinner:** Helena Andrews-Dyer and Emily Heil, "Japan State Dinner: The Toasts; Michelle Obama's Dress; Russell Wilson and Ciara Make a Public Appearance," **Washington Post,** April 28, 2015.

393 **Over at Fortune:** Roger Parloff, "Disruptive Diagnostics Firm Theranos Gets Boost from FDA," Fortune.com, July 2, 2015.

396 **Two months earlier, Balwani had terrorized:** Anonymous review of Theranos posted on Glassdoor.com on May 11, 2015.

399 **During the roundtable discussion:** Theranos, "Theranos Hosts Vice President Biden for Summit on a New Era of Preventive Health Care," press release, July 23, 2015, Theranos website.

399 **He also praised Holmes:** Ibid.

400 **A few days later, on July 28, I opened:** Elizabeth Holmes, "How to Usher in a New Era of

Preventive Health Care," **Wall Street Journal,** July 28, 2015.

23. DAMAGE CONTROL

403 **In March, a month after:** VC Experts report on Theranos Inc.

403 **Of the more than $430 million:** Christopher Weaver and John Carreyrou, "Theranos Offers Shares for Promise Not to Sue," **Wall Street Journal,** March 23, 2017.

404 **It was created by the Russian:** Breakthrough Prize website; https://breakthroughprize.org.

404 **Its cover letter stated:** Letter written by Elizabeth Holmes to Rupert Murdoch on Theranos letterhead dated December 4, 2014.

405 **The one call he placed:** Theranos announced an alliance with the Cleveland Clinic on March 9, 2017, in a press release titled "Theranos and Cleveland Clinic Announce Strategic Alliance to Improve Patient Care Through Innovation in Laboratory Testing," Theranos website.

405 **The investment packet she sent:** The projections were in a five-page document summarizing Theranos's financial situation, including information about its capitalization, cash flow, and balance sheet. They were first disclosed in Christopher Weaver and John Carreyrou, "Theranos Foresaw Huge Growth in Revenue and Profits," **Wall Street Journal,** December 5, 2016.

405 **They included Cox Enterprises:** Ibid.

406 **By the time Mike Siconolfi and I:** Altogether, Holmes had six meetings with Murdoch. They took place on November 26, 2014; April 22, 2015; July 3, 2015; September 29, 2015; January 30, 2016; and June 8, 2016. Two were in California and four in New York.

407 **Boies Schiller's Mike Brille:** Letter from Michael A. Brille to Mary L. Symons, Rochelle Gibbons's estate lawyer, dated August 5, 2015.

408 **In a last-ditch effort to prevent:** Letter from David Boies to Gerard Baker, copying Jason Conti, dated September 8, 2015.

411 **The story was published on:** John Carreyrou, "A Prized Startup's Struggles," **Wall Street Journal,** October 15, 2015.

411 **The editor of Fortune: Fortune CEO Daily** newsletter sent by Alan Murray to readers at 7:18 a.m. EST on October 15, 2015.

411 **Forbes and The New Yorker:** Matthew Herper, "Theranos' Elizabeth Holmes Needs to Stop Complaining and Answer Questions," Forbes.com, October 15, 2015; Eric Lach, "The Secrets of a Billionaire's Blood-Testing Startup," NewYorker.com, October 16, 2015.

411 **One of them was former Netscape cofounder:** Laura Arrillaga-Andreessen, "Five Visionary Tech Entrepreneurs Who Are Changing the World," **New York Times T Magazine,** October 12, 2015.

412 **In a press release it posted:** Theranos, "State-

ment from Theranos," press release, October 15, 2015, Theranos website.

414 **Dressed in her usual all-black attire:** Holmes's October 15, 2015, interview with Jim Cramer on CNBC's **Mad Money** program can be viewed on YouTube; https://www.youtube.com/watch?v=rGfaJZAdfNE.

414 **We quickly published my follow-up:** John Carreyrou, "Hot Startup Theranos Dials Back Lab Tests at FDA's Behest," **Wall Street Journal,** October 16, 2015.

415 **Theranos had issued a second:** Theranos, "Statement from Theranos," press release, October 16, 2015, Theranos website.

416 **At his signal:** Nick Bilton, "How Elizabeth Holmes's House of Cards Came Tumbling Down," **Vanity Fair,** September 6, 2016.

417 **There was so much interest:** Jonathan Krim's October 21, 2016, interview of Holmes at the WSJ D.Live conference can be viewed on WSJ.com.

419 **A few days earlier, Gassée:** Jean-Louis Gassée, "Theranos Trouble: A First Person Account," **Monday Note,** October 18, 2015.

419 **Soon after the interview ended:** Theranos, "Theranos Facts," press release, October 21, 2015, Theranos website.

420 **After Holmes's appearance:** Andrew Pollack, "Theranos, Facing Criticism, Says It Has Changed Board Structure," **New York Times,** October 28, 2015.

420 **Sure enough, within days:** Letters from Heather King to William Lewis, CEO of **Wall Street Journal** parent company Dow Jones, copying Mark Jackson, Jason Conti, Gerard Baker, John Carreyrou, and Mike Siconolfi, dated November 4 and 5, 2015.

420 **A third letter followed demanding:** Letter from Heather King to Jason Conti dated November 11, 2015.

420 **In an interview with Wired:** Nick Stockton, "The Theranos Scandal Could Become a Legal Nightmare," **Wired,** October 29, 2015.

421 **They revealed that Walgreens:** Michael Siconolfi, John Carreyrou, and Christopher Weaver, "Walgreens Scrutinizes Theranos Testing," **Wall Street Journal,** October 23, 2015.

421 **that Theranos had tried to sell:** Rolfe Winkler and John Carreyrou, "Theranos Authorizes New Shares That Could Raise Valuation," **Wall Street Journal,** October 28, 2015.

421 **that its lab was operating without:** John Carreyrou, "Theranos Searches for Director to Oversee Laboratory," **Wall Street Journal,** November 5, 2015.

421 **and that Safeway had walked away:** John Carreyrou, "Safeway, Theranos Split After $350 Million Deal Fizzles," **Wall Street Journal,** November 10, 2015.

421 **With each new story:** Letter from Heather King to William Lewis dated November 11, 2015.

421 **In an interview with Bloomberg Business-**

week: Sheelah Kolhatkar and Caroline Chen, "Can Elizabeth Holmes Save Her Unicorn?" **Bloomberg Businessweek,** December 10, 2015.

422 **In her acceptance speech:** Anne Cohen, "Reese Witherspoon Asks 'What Do We Do Now?' at **Glamour**'s Women of the Year Awards," **Variety,** November 9, 2015.

24. THE EMPRESS HAS NO CLOTHES

423 **Under the subject line:** Email with the subject line "CMS Complaint: Theranos Inc." sent by Erika Cheung to Gary Yamamoto at 6:13 p.m. PST on September 19, 2015.

426 **In late January, we were finally able:** John Carreyrou, Christopher Weaver, and Mike Siconolfi, "Deficiencies Found at Theranos Lab," **Wall Street Journal,** January 24, 2016.

426 **How serious became clear:** January 25, 2016, letter from Centers for Medicare and Medicaid Services official Karen Fuller to Theranos laboratory director Sunil Dhawan.

427 **Suddenly, Heather King's:** The last letter demanding retractions the **Wall Street Journal** received from Theranos is dated January 11, 2016.

427 **However, Theranos continued to minimize:** Email with the subject line "Statement by Theranos on CMS Audit Results" sent by Theranos spokeswoman Brooke Buchanan to journalists at 1:49 p.m. EST on January 27, 2016.

428 **the lab had continued to run:** John Carreyrou

and Christopher Weaver, "Theranos Ran Tests Despite Quality Problems," **Wall Street Journal,** March 8, 2016.

428 **Theranos couldn't refute:** Email with the subject line "statements from Theranos" sent by Brooke Buchanan to John Carreyrou and Mike Siconolfi at 3:35 p.m. EST March 7, 2016.

428 **But Heather King continued:** King sent CMS several letters in March and early April 2016 demanding that the agency make redactions before releasing the inspection report to the press.

429 **As the tug-of-war:** Noah Kulwin, "Theranos CEO Elizabeth Holmes Is Holding a Hillary Fundraiser with Chelsea Clinton," **Recode,** March 14, 2016.

429 **The fund-raiser was later relocated:** Ed Silverman, "Avoiding 'Teapot Tempest,' Clinton Campaign Distances Itself from Theranos," **STAT,** March 21, 2016.

430 **Heather King tried to prevent us:** Letter from Heather King to Jason Conti, copying John Carreyrou, Mike Siconolfi, and Gerard Baker, dated March 30, 2016.

430 **We posted it on the Journal's website:** John Carreyrou and Christopher Weaver, "Theranos Devices Often Failed Accuracy Requirements," **Wall Street Journal,** March 31, 2016.

430 **The coup de grâce:** Letter from CMS's Karen Fuller to Sunil Dhawan, Elizabeth Holmes, and Ramesh Balwani dated March 18, 2016.

431 **When we reported news of:** John Carreyrou

and Christopher Weaver, "Regulators Propose Banning Theranos Founder Elizabeth Holmes for at Least Two Years," **Wall Street Journal,** April 13, 2016.

431 **She had to come out:** Holmes's interview with Maria Shriver aired on April 18, 2016, and can be viewed on YouTube.

433 **In a complete about-face:** The AACC put out a press release on April 18, 2016, saying Holmes would present her technology at its sixty-eighth annual meeting.

434 **She broke up with him:** John Carreyrou, "Theranos Executive Sunny Balwani to Depart Amid Regulatory Probes," **Wall Street Journal,** May 12, 2016.

435 **A week later, we reported that:** John Carreyrou, "Theranos Voids Two Years of Edison Blood-Test Results," **Wall Street Journal,** May 18, 2016.

435 **On June 12, 2016, it terminated:** Michael Siconolfi, Christopher Weaver, and John Carreyrou, "Walgreen Terminates Partnership with Blood-Testing Firm Theranos," **Wall Street Journal,** June 13, 2016.

435 **In another crippling blow:** John Carreyrou, Michael Siconolfi, and Christopher Weaver, "Theranos Dealt Sharp Blow as Elizabeth Holmes Is Banned from Operating Labs," **Wall Street Journal,** July 8, 2016.

435 **More ominously, Theranos was now:** Christopher Weaver, John Carreyrou, and Michael

Siconolfi, "Theranos Is Subject of Criminal Probe by U.S.," **Wall Street Journal,** April 18, 2016.

436 **Over the next hour, Holmes proceeded:** Holmes's AACC presentation can be viewed on the association's website, AACC.org.

437 **While Holmes's presentation included:** The slides from Holmes's AACC presentation are available on AACC.org.

439 **A headline in Wired captured:** Nick Stockton, "Theranos Had a Chance to Clear Its Name. Instead, It Tried to Pivot," Wired.com, August 2, 2016.

439 **In an interview with the Financial Times:** David Crow, "Theranos Founder's Conference Invitation Sparks Row Among Scientists," **Financial Times,** August 4, 2016.

440 **But in another embarrassing setback:** John Carreyrou and Christopher Weaver, "Theranos Halts New Zika Test After FDA Inspection," **Wall Street Journal,** August 30, 2016.

440 **Partner Fund, the San Francisco hedge fund:** Christopher Weaver, "Major Investor Sues Theranos," **Wall Street Journal,** October 10, 2016.

440 **Another set of investors:** Christopher Weaver, "Theranos Sued for Alleged Fraud by Robertson Stephens Co-Founder Colman," **Wall Street Journal,** November 28, 2016.

440 **Most of the other investors opted:** Christopher Weaver and John Carreyrou, "Theranos Offers

Shares for Promise Not to Sue," **Wall Street Journal,** March 23, 2017.

440 **The media mogul sold his stock:** Ibid.

440 **David Boies and his law firm:** John Carreyrou, "Theranos and David Boies Cut Legal Ties," **Wall Street Journal,** November 20, 2016.

441 **A month after Holmes's AACC appearance:** Carreyrou and Weaver, "Theranos Halts New Zika Test After FDA Inspection."

441 **Boies left the Theranos board:** Weaver and Carreyrou, "Theranos Offers Shares for Promise Not to Sue."

441 **Walgreens, which had sunk:** Christopher Weaver, John Carreyrou, and Michael Siconolfi, "Walgreen Sues Theranos, Seeks $140 Million in Damages," **Wall Street Journal,** November 8, 2016.

441 **After initially attempting to appeal:** John Carreyrou and Christopher Weaver, "Theranos Retreats from Blood Tests," **Wall Street Journal,** October 6, 2016.

441 **During an inspection of the Arizona facility:** Christopher Weaver and John Carreyrou, "Second Theranos Lab Failed U.S. Inspection," **Wall Street Journal,** January 17, 2017.

441 **Under a settlement with Arizona's attorney general:** Christopher Weaver, "Arizona Attorney General Reaches Settlement with Theranos," **Wall Street Journal,** April 18, 2017.

441 **The number of test results:** Ibid.

Index

ABOUT THE AUTHOR

JOHN CARREYROU is a two-time Pulitzer Prize–winning investigative reporter at the **Wall Street Journal**. For his extensive coverage of Theranos, Carreyrou was awarded the George Polk Award for Financial Reporting, the Gerald Loeb Award for Distinguished Business and Financial Journalism in the category of beat reporting, and the Barlett & Steele Silver Award for Investigative Business Journalism. Carreyrou lives in Brooklyn with his wife and three children.